MATHEMATICAL MONOGRAPHS.

EDITED BY

MANSFIELD MERRIMAN AND ROBERT S. WOODWARD.

No. 4.

HYPERBOLIC FUNCTIONS.

BY

JAMES McMAHON,

PROFESSOR OF MATHEMATICS IN CORNELL UNIVERSITY.

NEW YORK:

JOHN WILEY & SONS.

LONDON: CHAPMAN & HALL, LIMITED.

1906.

COPYRIGHT 1896
BY
MANSFIELD MERRIMAN AND ROBERT S. WOODWARD
UNDER THE TITLE
HIGHER MATHEMATICS

TRANSCRIBER'S NOTE: *I did my best to recreate the index.*

MATHEMATICAL MONOGRAPHS.
EDITED BY
Mansfield Merriman and Robert S. Woodward.
Octavo. Cloth.

No. 1. History of Modern Mathematics.
By DAVID EUGENE SMITH. $1.00 *net*.

No. 2. Synthetic Projective Geometry.
By GEORGE BRUCE HALSTED. $1.00 *net*.

No. 3. Determinants.
By LAENAS GIFFORD WELD. $1.00 *net*.

No. 4. Hyperbolic Functions.
By JAMES MCMAHON. $1.00 *net*.

No. 5. Harmonic Functions.
By WILLIAM E. BYERLY. $1.00 *net*.

No. 6. Grassmann's Space Analysis.
By EDWARD W. HYDE. $1.00 *net*.

No. 7. Probability and Theory of Errors.
By ROBERT S. WOODWARD. $1.00 *net*.

No. 8. Vector Analysis and Quaternions.
By ALEXANDER MACFARLANE. $1.00 *net*.

No. 9. Differential Equations.
By WILLIAM WOOLSEY JOHNSON. $1.00 *net*.

No. 10. The Solution of Equations.
By MANSFIELD MERRIMAN. $1.00 *net*.

No. 11. Functions of a Complex Variable.
By THOMAS S. FISKE. $1.00 *net*.

No. 12. The Theory of Relativity.
By ROBERT D. CARMICHAEL. $1.00 *net*.

No. 13. The Theory of Numbers.
By ROBERT D. CARMICHAEL. $1.00 *net*.

No. 14. Algebraic Invariants.
By LEONARD E. DICKSON. $1.25 *net*.

PUBLISHED BY
JOHN WILEY & SONS, Inc., NEW YORK.
CHAPMAN & HALL, Limited, LONDON.

Editors' Preface.

The volume called Higher Mathematics, the first edition of which was published in 1896, contained eleven chapters by eleven authors, each chapter being independent of the others, but all supposing the reader to have at least a mathematical training equivalent to that given in classical and engineering colleges. The publication of that volume is now discontinued and the chapters are issued in separate form. In these reissues it will generally be found that the monographs are enlarged by additional articles or appendices which either amplify the former presentation or record recent advances. This plan of publication has been arranged in order to meet the demand of teachers and the convenience of classes, but it is also thought that it may prove advantageous to readers in special lines of mathematical literature.

It is the intention of the publishers and editors to add other monographs to the series from time to time, if the call for the same seems to warrant it. Among the topics which are under consideration are those of elliptic functions, the theory of numbers, the group theory, the calculus of variations, and non-Euclidean geometry; possibly also monographs on branches of astronomy, mechanics, and mathematical physics may be included. It is the hope of the editors that this form of publication may tend to promote mathematical study and research over a wider field than that which the former volume has occupied.

December, 1905.

Author's Preface.

This compendium of hyperbolic trigonometry was first published as a chapter in Merriman and Woodward's Higher Mathematics. There is reason to believe that it supplies a need, being adapted to two or three different types of readers. College students who have had elementary courses in trigonometry, analytic geometry, and differential and integral calculus, and who wish to know something of the hyperbolic trigonometry on account of its important and historic relations to each of those branches, will, it is hoped, find these relations presented in a simple and comprehensive way in the first half of the work. Readers who have some interest in imaginaries are then introduced to the more general trigonometry of the complex plane, where the circular and hyperbolic functions merge into one class of transcendents, the singly periodic functions, having either a real or a pure imaginary period. For those who also wish to view the subject in some of its practical relations, numerous applications have been selected so as to illustrate the various parts of the theory, and to show its use to the physicist and engineer, appropriate numerical tables being supplied for these purposes.

With all these things in mind, much thought has been given to the mode of approaching the subject, and to the presentation of fundamental notions, and it is hoped that some improvements are discernible. For instance, it has been customary to define the hyperbolic functions in relation to a sector of the rectangular hyperbola, and to take the initial radius of the sector coincident with the principal radius of the curve; in the present work, these and similar restrictions are discarded in the interest of analogy and generality, with a gain in symmetry and simplicity, and the functions are defined as certain characteristic ratios belonging to any sector of any hyperbola. Such definitions, in connection with the fruitful notion of correspondence of points on conics, lead to simple and general proofs of the addition-theorems, from which easily follow the conversion-formulas, the derivatives, the Maclaurin expansions, and the exponential expressions. The proofs are so arranged as to apply equally to the circular functions, regarded as the characteristic ratios belonging to any elliptic sector. For those, however, who may wish to start with the exponential expressions as the definitions of the hyperbolic functions, the appropriate order of procedure is indicated on page 27, and a direct mode of bringing such exponential definitions into geometrical relation with the hyperbolic sector is shown in the Appendix.

December, 1905.

Contents

Editors' Preface.	iii
Author's Preface.	iv
1 Correspondence of Points on Conics.	1
2 Areas of Corresponding Triangles.	3
3 Areas of Corresponding Sectors.	4
4 Charactersitic Ratios of Sectorial Measures.	5
5 Ratios Expressed as Triangle-measures.	6
6 Functional Relations for Ellipse.	7
7 Functional Relations for Hyperbola.	8
8 Relations Among Hyperbolic Functions.	9
9 Variations of the Hyperbolic Functions.	12
10 Anti-hyperbolic Functions.	14
11 Functions of Sums and Differences.	15
12 Conversion Formulas.	18
13 Limiting Ratios.	19
14 Derivatives of Hyperbolic Functions.	20
15 Derivatives of Anti-hyperbolic Functions.	23
16 Expansion of Hyperbolic Functions.	25
17 Exponential Expressions.	27

CONTENTS

18 Expansion of Anti-functions. — 29

19 Logarithmic Expression of Anti-Functions. — 31

20 The Gudermanian Function. — 33

21 Circular Functions of Gudermanian. — 34

22 Gudermanian Angle — 36

23 Derivatives of Gudermanian and Inverse. — 38

24 Series for Gudermanian and its Inverse. — 40

25 Graphs of Hyperbolic Functions. — 42

26 Elementary Integrals. — 45

27 Functions of Complex Numbers. — 49

28 Addition-Theorems for Complexes. — 51

29 Functions of Pure Imaginaries. — 53

30 Functions of $x + iy$ in the Form $X + iY$. — 55

31 The Catenary — 59

32 Catenary of Uniform Strength. — 61

33 The Elastic Catenary. — 63

34 The Tractory. — 65

35 The Loxodrome. — 67

36 Combined Flexure and Tension. — 69

37 Alternating Currents. — 71

38 Miscellaneous Applications. — 77

39 Explanation of Tables. — 79

40 Appendix. — 88
   40.1 Historical and Bibliographical. — 88
   40.2 Exponential Expressions as Definitions. — 89

Index — 89

Article 1

Correspondence of Points on Conics.

To prepare the way for a general treatment of the hyperbolic functions a preliminary discussion is given on the relations, between hyperbolic sectors. The method adopted is such as to apply at the same time to sectors of the ellipse, including the circle; and the analogy of the hyperbolic and circular functions will be obvious at every step, since the same set of equations can be read in connection with either the hyperbola or the ellipse.[1] It is convenient to begin with the theory of correspondence of points on two central conics of like species, i.e. either both ellipses or both hyperbolas.

---

[1] The hyperbolic functions are not so named on account of any analogy with what are termed Elliptic Functions. "The elliptic integrals, and thence the elliptic functions, derive their name from the early attempts of mathematicians at the rectification of the ellipse.... To a certain extent this is a disadvantage; ... because we employ the name hyperbolic function to denote $\cosh u, \sinh u$, etc., by analogy with which the elliptic functions would be merely the circular functions $\cos \phi, \sin \phi$, etc...." (Greenhill, Elliptic Functions, p. 175.)

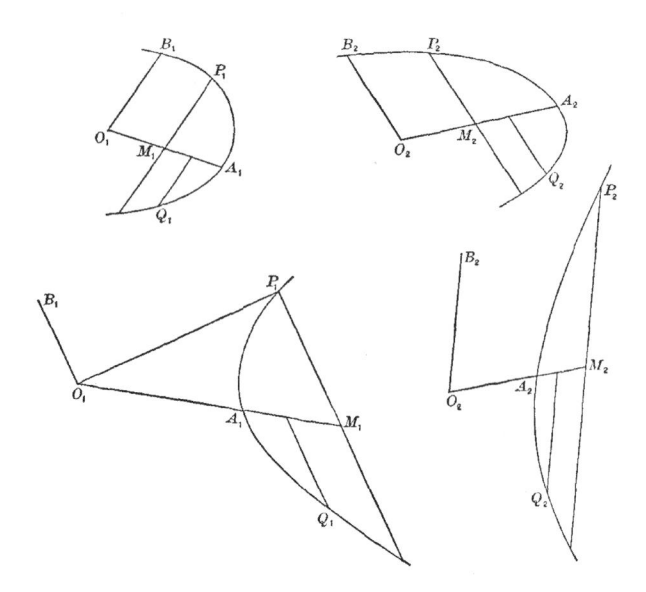

To obtain a definition of corresponding points, let O_1A_1, O_1B_1 be conjugate radii of a central conic, and O_2A_2, O_2B_2 conjugate radii of any other central conic of the same species; let P_1, P_2 be two points on the curves; and let their coordinates referred to the respective pairs of conjugate directions be $(x_1, y_1), (x_2, y_2)$; then, by analytic geometry,

$$\frac{x_1^2}{a_1^2} \pm \frac{y_1^2}{b_1^2} = 1, \qquad \frac{x_2^2}{a_2^2} \pm \frac{y_2^2}{b_2^2} = 1. \tag{1}$$

Now if the points P_1, P_2 be so situated that

$$\frac{x_1}{a_1} = \frac{x_2}{a_2}, \qquad \frac{y_1}{b_1} = \frac{y_2}{b_2}, \tag{2}$$

the equalities referring to sign as well as magnitude, then P_1, P_2 are called corresponding points in the two systems. If Q_1, Q_2 be another pair of correspondents, then the sector and triangle $P_1O_1Q_1$ are said to correspond respectively with the sector and triangle $P_2O_2Q_2$. These definitions will apply also when the conics coincide, the points P_1, P_2 being then referred to any two pairs of conjugate diameters of the same conic.

In discussing the relations between corresponding areas it is convenient to adopt the following use of the word "measure": The measure of any area connected with a given central conic is the ratio which it bears to the constant area of the triangle formed by two conjugate diameters of the same conic.

For example, the measure of the sector $A_1O_1P_1$ is the ratio

$$\frac{\text{sector } A_1O_1P_1}{\text{triangle } A_1O_1B_1}$$

and is to be regarded as positive or negative according as $A_1O_1P_1$ and $A_1O_1B_1$ are at the same or opposite sides of their common initial line.

Article 2

Areas of Corresponding Triangles.

The areas of corresponding triangles have equal measures. For, let the coordinates of P_1, Q_1 be $(x_1, y_1), (x_1', y_1')$, and let those of their correspondents P_2, Q_2 be $(x_2, y_2), (x_2', y_2')$; let the triangles $P_1 O_1 Q_1, P_2 O_2 Q_2$ be T_1, T_2, and let the measuring triangles $A_1 O_1 B_1, A_2 O_2 B_2$ be K_1, K_2, and their angles ω_1, ω_1; then, by analytic geometry, taking account of both magnitude and direction of angles, areas, and lines,

$$\frac{T_1}{K_1} = \frac{\frac{1}{2}(x_1 y_1' - x_1' y_1) \sin \omega_1}{\frac{1}{2} a_1 b_1 \sin \omega_1} = \frac{x_1}{a_1} \cdot \frac{y_1'}{b_1} - \frac{x_1'}{a_1} \cdot \frac{y_1}{b_1};$$

$$\frac{T_2}{K_2} = \frac{\frac{1}{2}(x_2 y_2' - x_2' y_2) \sin \omega_2}{\frac{1}{2} a_2 b_2 \sin \omega_2} = \frac{x_2}{a_2} \cdot \frac{y_2'}{b_2} - \frac{x_2'}{a_2} \cdot \frac{y_2}{b_2}.$$

Therefore, by (2),

$$\frac{T_1}{K_1} = \frac{T_2}{K_2}. \tag{3}$$

Article 3

Areas of Corresponding Sectors.

The areas of corresponding sectors have equal measures. For conceive the sectors S_1, S_2 divided up into infinitesimal corresponding sectors; then the respective infinitesimal corresponding triangles have equal measures (Art. 2); but the given sectors are the limits of the sums of these infinitesimal triangles, hence

$$\frac{S_1}{K_1} = \frac{S_2}{K_2}. \tag{4}$$

In particular, the sectors $A_1 O_1 P_1, A_2 O_2 P_2$ have equal measures; for the initial points A_1, A_2 are corresponding points.

It may be proved conversely by an obvious reductio ad absurdum that if the initial points of two equal-measured sectors correspond, then their terminal points correspond.

Thus if any radii $O_1 A_1, O_2 A_2$ be the initial lines of two equal-measured sectors whose terminal radii are $O_1 P_1, O_2 P_2$, then P_1, P_2 are corresponding points referred respectively to the pairs of conjugate directions $O_1 A_1, O_1 B_1$, and $O_2 A_2, O_2 A_B$; that is,

$$\frac{x_1}{a_1} = \frac{x_2}{a_2}, \quad \frac{y_1}{b_1} = \frac{y_2}{b_2}.$$

Prob. 1. Prove that the sector $P_1 O_1 Q_1$, is bisected by the line joining O_1, to the mid-point of $P_1 Q_1$. (Refer the points P_1, Q_1, respectively, to the median as common axis of x, and to the two opposite conjugate directions as axis of y, and show that P_1, Q_1 are then corresponding points.)

Prob. 2. Prove that the measure of a circular sector is equal to the radian measure of its angle.

Prob. 3. Find the measure of an elliptic quadrant, and of the sector included by conjugate radii.

Article 4

Charactersitic Ratios of Sectorial Measures.

Let $A_1 O_1 P_1 = S_1$, be any sector of a central conic; draw $P_1 M_1$ ordinate to $O_1 A_1$, i.e. parallel to the tangent at A_1; let $O_1 M_1 = x_1, M_1 P_1 = y_1, O_1 A_1 = a_1$, and the conjugate radius $O_1 B_1 = b_1$; then the ratios $\dfrac{x_1}{a_1}, \dfrac{y_1}{b_1}$ are called the characteristic ratios of the given sectorial measure $\dfrac{S_1}{K_1}$. These ratios are constant both in magnitude and sign for all sectors of the same measure and species wherever these may be situated (Art. 3). Hence there exists a functional relation between the sectorial measure and each of its characteristic ratios.

Article 5

Ratios Expressed as Triangle-measures.

The triangle of a sector and its complementary triangle are measured by the two characteristic ratios. For, let the triangle $A_1 O_1 P_1$ and its complementary triangle $P_1 O_1 B_1$ be denoted by T_1, T_1'; then

$$\left. \begin{array}{l} \dfrac{T_1}{K_1} = \dfrac{\frac{1}{2} a_1 y_1 \sin \omega_1}{\frac{1}{2} a_1 b_1 \sin \omega_1} = \dfrac{y_1}{b_1}, \\[2mm] \dfrac{T_1'}{K_1} = \dfrac{\frac{1}{2} b_1 x_1 \sin \omega_1}{\frac{1}{2} a_1 b_1 \sin \omega_1} = \dfrac{x_1}{a_1}. \end{array} \right\} \quad (5)$$

Article 6

Functional Relations for Ellipse.

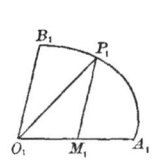

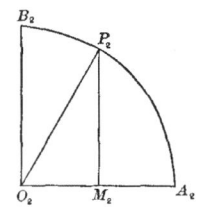

The functional relations that exist between the sectorial measure and each of its characteristic ratios are the same for all elliptic, including circular, sectors (Art. 4). Let P_1, P_2 be corresponding points on an ellipse and a circle, referred to the conjugate directions O_1A_1, O_1B_1 and O_2A_2, O_2B_2, the latter pair being at right angles; let the angle $A_2O_2P_2 = \theta$ in radian measure; then

$$\frac{S_2}{K_2} = \frac{\frac{1}{2}a_2^2\theta}{\frac{1}{2}a_2^2} = \theta. \tag{6}$$

$$\therefore \frac{x_2}{a_2} = \cos\frac{S_2}{K_2}, \quad \frac{y_2}{b_2} = \sin\frac{S_2}{K_2}; \quad [a_2 = b_2]$$

hence, in the ellipse, by Art. 3,

$$\frac{x_1}{a_1} = \cos\frac{S_1}{K_1}, \quad \frac{y_1}{b_1} = \sin\frac{S_1}{K_1}. \tag{7}$$

Prob. 4. Given $x_1 = \frac{1}{2}a_1$; find the measure of the elliptic sector $A_1O_1P_1$. Also find its area when $a_1 = 4, b_1 = 3, \omega = 60°$.

Prob. 5. Find the characteristic ratios of an elliptic sector whose measure is $\frac{1}{4}\pi$.

Prob. 6. Write down the relation between an elliptic sector and its triangle. (See Art. 5.)

Article 7

Functional Relations for Hyperbola.

The functional relations between a sectorial measure and its characteristic ratios in the case of the hyperbola may be written in the form

$$\frac{x_1}{a_1} = \cosh \frac{S_1}{K_1}, \quad \frac{y_1}{b_1} = \sinh \frac{S_1}{K_1},$$

and these express that the ratio of the two lines on the left is a certain definite function of the ratio of the two areas on the right. These functions are called by analogy the hyperbolic cosine and the hyperbolic sine. Thus, writing u for $\frac{S_1}{K_1}$ the two equations

$$\frac{x_1}{a_1} = \cosh u, \quad \frac{y_1}{b_1} = \sinh u \qquad (8)$$

serve to define the hyperbolic cosine and sine of a given sectorial measure u; and the hyperbolic tangent, cotangent, secant, and cosecant are then defined as follows:

$$\left. \begin{array}{ll} \tanh u = \dfrac{\sinh u}{\cosh u}, & \coth u = \dfrac{\cosh u}{\sinh u}, \\[2mm] \operatorname{sech} u = \dfrac{1}{\cosh u}, & \operatorname{csch} u = \dfrac{1}{\sinh u}. \end{array} \right\} \qquad (9)$$

The names of these functions may be read "h-cosine," "h-sine," "h-tangent," etc., or "hyper-cosine," etc.

Article 8

Relations Among Hyperbolic Functions.

Among the six functions there are five independent relations, so that when the numerical value of one of the functions is given, the values of the other five can be found. Four of these relations consist of the four defining equations (9). The fifth is derived from the equation of the hyperbola

$$\frac{x_1^2}{a_1^2} - \frac{y_1^2}{b_1^2} = 1,$$

giving

$$\cosh^2 u - \sinh^2 u = 1. \tag{10}$$

By a combination of some of these equations other subsidiary relations may be obtained; thus, dividing (10) successively by $\cosh^2 u, \sinh^2 u$, and applying (9), give

$$\left.\begin{array}{r}1 - \tanh^2 u = \operatorname{sech}^2 u, \\ \coth^2 u - 1 = \operatorname{csch}^2 u.\end{array}\right\} \tag{11}$$

Equations (9), (10), (11) will readily serve to express the value of any function in terms of any other. For example, when $\tanh u$ is given,

$$\coth u = \frac{1}{\tanh u}, \quad \operatorname{sech} u = \sqrt{1 - \tanh^2 u},$$

$$\cosh u = \frac{1}{\sqrt{1 - \tanh^2 u}}, \quad \sinh u = \frac{\tanh u}{\sqrt{1 - \tanh^2 u}},$$

$$\operatorname{csch} u = \frac{\sqrt{1 - \tanh^2 u}}{\tanh u}.$$

The ambiguity in the sign of the square root may usually be removed by the following considerations: The functions $\cosh u, \operatorname{sech} u$ are always positive,

ARTICLE 8. RELATIONS AMONG HYPERBOLIC FUNCTIONS.

because the primary characteristic ratio $\dfrac{x_1}{a_1}$ is positive, since the initial line $O_1 A_1$ and the abscissa $O_1 M_1$ are similarly directed from O_1 on whichever branch of the hyperbola P_1 maybe situated; but the functions $\sinh u, \tanh u, \coth u, \operatorname{csch} u$, involve the other characteristic ratio $\dfrac{y_1}{b_1}$, which is positive or negative according as y_1 and b_1 have the same or opposite signs, i.e., as the measure u is positive or negative; hence these four functions are either all positive or all negative. Thus when any one of the functions $\sinh u, \tanh u, \operatorname{csch} u, \coth u$, is given in magnitude and sign, there is no ambiguity in the value of any of the six hyperbolic functions; but when either $\cosh u$ or $\operatorname{sech} u$ is given, there is ambiguity as to whether the other four functions shall be all positive or all negative.

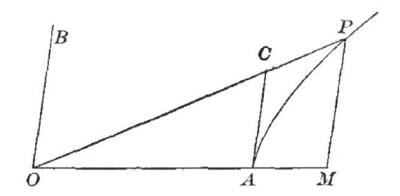

The hyperbolic tangent may be expressed as the ratio of two lines. For draw the tangent line $AC = t$; then

$$\tanh u = \frac{y}{b} : \frac{x}{a} = \frac{a}{b} \cdot \frac{y}{x}$$
$$= \frac{a}{b} \cdot \frac{t}{a} = \frac{t}{b}. \tag{12}$$

The hyperbolic tangent is the measure of the triangle OAC. For

$$\frac{OAC}{OAB} = \frac{at}{ab} = \frac{t}{b} = \tanh u. \tag{13}$$

Thus the sector AOP, and the triangles AOP, POB, AOC, are proportional to $u, \sinh u, \cosh u, \tanh u$ (eqs. 5, 13); hence

$$\sinh u > u > \tanh u. \tag{14}$$

Prob. 7. Express all the hyperbolic functions in terms of $\sinh u$. Given $\cosh u = 2$, find the values of the other functions.

Prob. 8. Prove from eqs. 10, 11, that $\cosh u > \sinh u, \cosh u > 1, \tanh u < 1, \operatorname{sech} u < 1$.

Prob. 9. In the figure of Art. 1, let $OA = 2, OB = 1, AOB = 60°$, and area of sector $AOP = 3$; find the sectorial measure, and the two characteristic ratios, in the elliptic sector, and also in the hyperbolic sector; and find the area of the triangle AOP. (Use tables of cos, sin, cosh, sinh.)

Prob. 10. Show that $\coth u, \operatorname{sech} u, \operatorname{csch} u$ may each be expressed as the ratio of two lines, as follows: Let the tangent at P make on the conjugate axes OA, OB, intercepts $OS = m, OT = n$; let the tangent at B, to the conjugate hyperbola, meet OP in R, making $BR = l$; then

$$\coth u = \frac{l}{a}, \quad \operatorname{sech} u = \frac{m}{a}, \quad \operatorname{csch} u = \frac{n}{b}.$$

ARTICLE 8. RELATIONS AMONG HYPERBOLIC FUNCTIONS.

Prob. 11. The measure of segment AMP is $\sinh u \cosh u - u$. Modify this for the ellipse. Modify also eqs. 10–14, and probs. 8, 10.

Article 9

Variations of the Hyperbolic Functions.

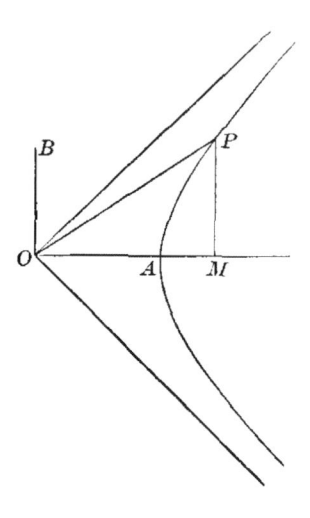

Since the values of the hyperbolic functions depend only on the sectorial measure, it is convenient, in tracing their variations, to consider only sectors of one half of a rectangular hyperbola, whose conjugate radii are equal, and to take the principal axis OA as the common initial line of all the sectors. The sectorial measure u assumes every value from $-\infty$, through 0, to $+\infty$, as the terminal point P comes in from infinity on the lower branch, and passes to infinity on the upper branch; that is, as the terminal line OP swings from the lower asymptotic position $y = -x$, to the upper one, $y = x$. It is here assumed, but is proved in Art. 17, that the sector AOP becomes infinite as P passes to infinity.

Since the functions $\cosh u, \sinh u, \tanh u$, for any position of OP, are equal to the ratios of x, y, t, to the principal radius a, it is evident from the figure that

$$\cosh 0 = 1, \quad \sinh 0 = 0, \quad \tanh 0 = 0, \qquad (15)$$

ARTICLE 9. VARIATIONS OF THE HYPERBOLIC FUNCTIONS.

and that as u increases towards positive infinity, $\cosh u, \sinh u$ are positive and become infinite, but $\tanh u$ approaches unity as a limit; thus

$$\cosh \infty = \infty, \quad \sinh \infty = \infty, \quad \tanh \infty = 1. \tag{16}$$

Again, as u changes from zero towards the negative side, $\cosh u$ is positive and increases from unity to infinity, but $\sinh u$ is negative and increases numerically from zero to a negative infinite, and $\tanh u$ is also negative and increases numerically from zero to negative unity; hence

$$\cosh(-\infty) = \infty, \quad \sinh(-\infty) = -\infty, \quad \tanh(-\infty) = -1. \tag{17}$$

For intermediate values of u the numerical values of these functions can be found from the formulas of Arts. 16, 17, and are tabulated at the end of this chapter. A general idea of their manner of variation can be obtained from the curves in Art. 25, in which the sectorial measure u is represented by the abscissa, and the values of the functions $\cosh u$, $\sinh u$, etc., are represented by the ordinate.

The relations between the functions of $-u$ and of u are evident from the definitions, as indicated above, and in Art. 8. Thus

$$\left.\begin{array}{ll} \cosh(-u) = +\cosh u, & \sinh(-u) = -\sinh u, \\ \operatorname{sech}(-u) = +\operatorname{sech} u, & \operatorname{csch}(-u) = -\operatorname{csch} u, \\ \tanh(-u) = -\tanh u, & \coth(-u) = -\coth u. \end{array}\right\} \tag{18}$$

Prob. 12. Trace the changes in $\operatorname{sech} u, \coth u, \operatorname{csch} u$, as u passes from $-\infty$ to $+\infty$. Show that $\sinh u, \cosh u$ are infinites of the same order when u is infinite. (It will appear in Art. 17 that $\sinh u, \cosh u$ are infinites of an order infinitely higher than the order of u.)

Prob. 13. Applying eq. (12) to figure, page 12, prove $\tanh u_1 = \tan AOP$.

Article 10

Anti-hyperbolic Functions.

The equations $\dfrac{x}{a} = \cosh u$, $\dfrac{y}{b} = \sinh u$, $\dfrac{t}{b} = \tanh u$, etc., may also be expressed by the inverse notation $u = \cosh^{-1}\dfrac{x}{a}$, $u = \sinh^{-1}\dfrac{y}{b}$, $u = \tanh^{-1}\dfrac{t}{b}$, etc., which may be read: "u is the sectorial measure whose hyperbolic cosine is the ratio x to a," etc.; or "u is the anti-h-cosine of $\dfrac{x}{a}$," etc.

Since there are two values of u, with opposite signs, that correspond to a given value of $\cosh u$, it follows that if u be determined from the equation $\cosh u = m$, where m is a given number greater than unity, u is a two-valued function of m. The symbol $\cosh^{-1} m$ will be used to denote the positive value of u that satisfies the equation $\cosh u = m$. Similarly the symbol $\operatorname{sech}^{-1} m$ in will stand for the positive value of u that satisfies the equation $\operatorname{sech} u = m$. The signs of the other functions $\sinh^{-1} m, \tanh^{-1} m, \coth^{-1} m, \operatorname{csch}^{-1} m$, are the same as the sign of m. Hence all of the anti-hyperbolic functions of real numbers are one-valued.

Prob. 14. Prove the following relations:

$$\cosh^{-1} m = \sinh^{-1}\sqrt{m^2-1}, \quad \sinh^{-1} m = \pm\cosh^{-1}\sqrt{m^2+1},$$

the upper or lower sign being used according as m is positive or negative. Modify these relations for $\sin^{-1}, \cos^{-1}$.

Prob. 15. In figure, Art. 1, let $OA = 2, OB = 1, AOB = 60°$; find the area of the hyperbolic sector AOP, and of the segment AMP, if the abscissa of P is 3. (Find $\cosh^{-1}$ from the tables for $\cosh$.)

Article 11

Functions of Sums and Differences.

(a) To prove the difference-formulas

$$\left.\begin{array}{l}\sinh(u-v) = \sinh u \cosh v - \cosh u \sinh v,\\ \cosh(u-v) = \cosh u \cosh v - \sinh u \sinh v.\end{array}\right\} \quad (19)$$

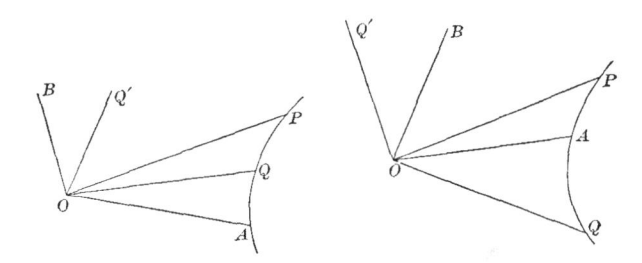

Let OA be any radius of a hyperbola, and let the sectors AOP, AOQ have the measures u, v; then $u-v$ is the measure of the sector QOP. Let OB, OQ' be the radii conjugate to OA, OQ; and let the coördinates of P, Q, Q' be (x_1, y_1), (x, y), (x', y') with reference to the axes OA, OB; then

$$\sinh(u-v) = \sinh \frac{\text{sector } QOP}{K} = \frac{\text{triangle } QOP}{K} \quad [\text{Art. 5.}]$$

$$= \frac{\frac{1}{2}(xy_1 - x_1 y)\sin\omega}{\frac{1}{2}a_1 b_1 \sin\omega} = \frac{y_1}{b_1}\cdot\frac{x}{a_1} - \frac{y}{b_1}\cdot\frac{x_1}{a_1}$$

$$= \sinh u \cosh v - \cosh u \sinh v;$$

ARTICLE 11. FUNCTIONS OF SUMS AND DIFFERENCES.

$$\cosh(u-v) = \cosh\frac{\text{sector } QOP}{K} = \frac{\text{triangle } POQ'}{K} \quad [\text{Art. 5.}$$

$$= \frac{\frac{1}{2}(x_1 y' - y_1 x')\sin\omega}{\frac{1}{2}a_1 b_1 \sin\omega} = \frac{y'}{b_1}\cdot\frac{x_1}{a_1} - \frac{y}{b_1}\cdot\frac{x'}{a_1};$$

but

$$\frac{y'}{b_1} = \frac{x}{a_1}, \quad \frac{x'}{a_1} = \frac{y}{b_1}, \tag{20}$$

since Q, Q' are extremities of conjugate radii; hence

$$\cosh(u-v) = \cosh u \cosh v - \sinh u \sinh v.$$

In the figures u is positive and v is positive or negative. Other figures may be drawn with u negative, and the language in the text will apply to all. In the case of elliptic sectors, similar figures may be drawn, and the same language will apply, except that the second equation of (20) will be $\frac{x'}{a_1} = \frac{-y}{b_1}$; therefore

$$\sin(u-v) = \sin u \cos v - \cos u \sin v,$$
$$\cos(u-v) = \cos u \cos v + \sin u \sin v.$$

(b) To prove the sum-formulas

$$\left.\begin{array}{l}\sinh(u+v) = \sinh u \cosh v + \cosh u \sinh v, \\ \cosh(u+v) = \cosh u \cosh v + \sinh u \sinh v.\end{array}\right\} \tag{21}$$

These equations follow from (19) by changing v into $-v$, and then for $\sinh(-v), \cosh(-v)$, writing $-\sinh v, \cosh v$ (Art. 9, eqs. (18)).

(c) To prove that

$$\tanh(u\pm v) = \frac{\tanh u \pm \tanh v}{1 \pm \tanh u \tanh v}. \tag{22}$$

Writing $\tanh(u\pm v) = \dfrac{\sinh(u\pm v)}{\cosh(u\pm v)}$, expanding and dividing numerator and denominator by $\cosh u \cosh v$, eq. (22) is obtained.

Prob. 16. Given $\cosh u = 2, \cosh v = 3$, find $\cosh(u+v)$.

Prob. 17. Prove the following identities:

(a) $\sinh 2u = 2\sinh u \cosh u.$

(b) $\cosh 2u = \cosh^2 u + \sinh^2 u = 1 + 2\sinh^2 u = 2\cosh^2 u - 1.$

(c) $1 + \cosh u = 2\cosh^2 \tfrac{1}{2}u, \cosh u - 1 = 2\sinh^2 \tfrac{1}{2}u.$

(d) $\tanh \tfrac{1}{2}u = \dfrac{\sinh u}{1+\cosh u} = \dfrac{\cosh u - 1}{\sinh u} = \left(\dfrac{\cosh u - 1}{\cosh u + 1}\right)^{\frac{1}{2}}.$

ARTICLE 11. FUNCTIONS OF SUMS AND DIFFERENCES.

(e) $\sinh 2u = \dfrac{2\tanh u}{1 - \tanh^2 u}$, $\cosh 2u = \dfrac{1 + \tanh^2 u}{1 - \tanh^2 u}$.

(f) $\sinh 3u = 3\sinh u + 4\sinh^3 u$, $\cosh 3u = 4\cosh^3 u - 3\cosh u$.

(g) $\cosh u + \sinh u = \dfrac{1 + \tanh \frac{1}{2}u}{1 - \tanh \frac{1}{2}u}$.

(h) $(\cosh u + \sinh u)(\cosh v + \sinh v) = \cosh(u+v) + \sinh(u+v)$.

(i) Generalize (h); and show also what it becomes when $u = v = \ldots$

(j) $\sinh^2 x \cos^2 y + \cosh^2 x \sin^2 y = \sinh^2 x + \sin^2 y$.

(k) $\cosh^{-1} m \pm \cosh^{-1} n = \cosh^{-1}\left[mn \pm \sqrt{(m^2-1)(n^2-1)}\right]$.

(l) $\sinh^{-1} m \pm \sinh^{-1} n = \sinh^{-1}\left[m\sqrt{1+n^2} \pm n\sqrt{1+m^2}\right]$.

Prob. 18. What modifications of signs are required in (21), (22), in order to pass to circular functions?

Prob. 19. Modify the identities of Prob. 17 for the same purpose.

Article 12

Conversion Formulas.

To prove that

$$\left.\begin{aligned}\cosh u_1 + \cosh u_2 &= 2\cosh\tfrac{1}{2}(u_1+u_2)\cosh\tfrac{1}{2}(u_1-u_2),\\ \cosh u_1 - \cosh u_2 &= 2\sinh\tfrac{1}{2}(u_1+u_2)\sinh\tfrac{1}{2}(u_1-u_2),\\ \sinh u_1 + \sinh u_2 &= 2\sinh\tfrac{1}{2}(u_1+u_2)\cosh\tfrac{1}{2}(u_1-u_2),\\ \sinh u_1 - \sinh u_2 &= 2\cosh\tfrac{1}{2}(u_1+u_2)\sinh\tfrac{1}{2}(u_1-u_2).\end{aligned}\right\} \quad (23)$$

From the addition formulas it follows that

$$\begin{aligned}\cosh(u+v)+\cosh(u-v)&=2\cosh u\cosh v,\\ \cosh(u+v)-\cosh(u-v)&=2\sinh u\sinh v,\\ \sinh(u+v)+\sinh(u-v)&=2\sinh u\cosh v,\\ \sinh(u+v)-\sinh(u-v)&=2\cosh u\sinh v,\end{aligned}$$

and then by writing $u+v = u_1$, $u-v = u_2$, $u = \tfrac{1}{2}(u_1+u_2)$, $v = \tfrac{1}{2}(u_1-u_2)$, these equations take the form required.

Prob. 20. In passing to circular functions, show that the only modification to be made in the conversion formulas is in the algebraic sign of the right-hand member of the second formula.

Prob. 21. Simplify $\dfrac{\cosh 2u + \cosh 4v}{\sinh 2u + \sinh 4v}$, $\dfrac{\cosh 2u + \cosh 4v}{\cosh 2u - \cosh 4v}$.

Prob. 22. Prove $\sinh^2 x - \sinh^2 y = \sinh(x+y)\sinh(x-y)$.

Prob. 23. Simplify $\cosh^2 x \cosh^2 y \pm \sinh^2 x \sinh^2 y$.

Prob. 24. Simplify $\cosh^2 x \cos^2 y + \sinh^2 x \sin^2 y$.

Article 13

Limiting Ratios.

To find the limit, as u approaches zero, of
$$\frac{\sinh u}{u}, \frac{\tanh u}{u},$$
which are then indeterminate in form.

By eq. (14), $\sinh u > u > \tanh u$; and if $\sinh u$ and $\tanh u$ be successively divided by each term of these inequalities, it follows that
$$1 < \frac{\sinh u}{u} < \cosh u,$$
$$\operatorname{sech} u < \frac{\tanh u}{u} < 1,$$
but when $u \doteq 0$, $\cosh u \doteq 1$, $\operatorname{sech} u \doteq 1$, hence
$$\lim_{u \doteq 0} \frac{\sinh u}{u} = 1, \lim_{u \doteq 0} \frac{\tanh u}{u} = 1. \tag{24}$$

Article 14

Derivatives of Hyperbolic Functions.

To prove that

$$\left.\begin{array}{rl}(a) & \dfrac{d(\sinh u)}{du} = \cosh u, \\ (b) & \dfrac{d(\cosh u)}{du} = \sinh u, \\ (c) & \dfrac{d(\tanh u)}{du} = \operatorname{sech}^2 u, \\ (d) & \dfrac{d(\operatorname{sech} u)}{du} = -\operatorname{sech} u \,\tanh u, \\ (e) & \dfrac{d(\coth u)}{du} = -\operatorname{csch}^2 u, \\ (f) & \dfrac{d(\operatorname{csch} u)}{du} = -\operatorname{csch} u \,\coth u, \end{array}\right\} \quad (25)$$

(a) Let

$$y = \sinh u,$$
$$\Delta y = \sinh(u + \Delta u) - \sinh u$$
$$= 2\cosh\tfrac{1}{2}(2u + \Delta u)\sinh\tfrac{1}{2}\Delta u,$$
$$\frac{\Delta y}{\Delta u} = \cosh\left(u + \tfrac{1}{2}\Delta u\right)\frac{\sinh \tfrac{1}{2}\Delta u}{\tfrac{1}{2}\Delta u}.$$

ARTICLE 14. DERIVATIVES OF HYPERBOLIC FUNCTIONS.

Take the limit of both sides, as $\Delta u \doteq 0$, and put

$$\lim. \frac{\Delta y}{\Delta u} = \frac{dy}{du} = \frac{d(\sinh u)}{du},$$

$$\lim. \cosh\left(u + \frac{1}{2}\Delta u\right) = \cosh u,$$

$$\lim. \frac{\sinh \frac{1}{2}\Delta u}{\frac{1}{2}\Delta u} = 1; \qquad \text{(see Art. 13)}$$

then

$$\frac{d(\sinh u)}{du} = \cosh u.$$

(b) Similar to (a).

(c)

$$\frac{d(\tanh u)}{du} = \frac{d}{du} \cdot \frac{\sinh u}{\cosh u}$$

$$= \frac{\cosh^2 u - \sinh^2 u}{\cosh^2 u} = \frac{1}{\cosh^2 u} = \operatorname{sech}^2 u.$$

(d) Similar to (c).

(e)

$$\frac{d(\operatorname{sech} u)}{du} = \frac{d}{du} \cdot \frac{1}{\cosh u} = -\frac{\sinh u}{\cosh^2 u} = -\operatorname{sech} u \tanh u.$$

(f) Similar to (e).

It thus appears that the functions $\sinh u$, $\cosh u$ reproduce themselves in two differentiations; and, similarly, that the circular functions $\sin u$, $\cos u$ produce their opposites in two differentiations. In this connection it may be noted that the frequent appearance of the hyperbolic (and circular) functions in the solution of physical problems is chiefly due to the fact that they answer the question: What function has its second derivative equal to a positive (or negative) constant multiple of the function itself? (See Probs. 28–30.) An answer such as $y = \cosh mx$ is not, however, to be understood as asserting that mx is an actual sectorial measure and y its characteristic ratio; but only that the relation between the numbers mx and y is the same as the known relation between the measure of a hyperbolic sector and its characteristic ratio; and that the numerical value of y could be found from a table of hyperbolic cosines.

Prob. 25. Show that for circular functions the only modifications required are in the algebraic signs of (b), (d).

ARTICLE 14. DERIVATIVES OF HYPERBOLIC FUNCTIONS.

Prob. 26. Show from their derivatives which of the hyperbolic and circular functions diminish as u increases.

Prob. 27. Find the derivative of $\tanh u$ independently of the derivatives of $\sinh u$, $\cosh u$.

Prob. 28. Eliminate the constants by differentiation from the equation
$$y = A \cosh mx + B \sinh mx,$$
and prove that $\dfrac{d^2y}{dx^2} = m^2 y.$

Prob. 29. Eliminate the constants from the equation
$$y = A \cos mx + B \sin mx,$$
and prove that $\dfrac{d^2y}{dx^2} = -m^2 y.$

Prob. 30. Write down the most general solutions of the differential equations
$$\frac{d^2y}{dx^2} = m^2 y, \quad \frac{d^2y}{dx^2} = -m^2 y, \quad \frac{d^4y}{dx^4} = m^4 y.$$

Article 15

Derivatives of Anti-hyperbolic Functions.

$$
\begin{aligned}
(a) \quad & \frac{d(\sinh^{-1} x)}{dx} = \frac{1}{\sqrt{x^2+1}}, \\
(b) \quad & \frac{d(\cosh^{-1} x)}{dx} = \frac{1}{\sqrt{x^2-1}}, \\
(c) \quad & \frac{d(\tanh^{-1} x)}{dx} = \frac{1}{1-x^2}\bigg]_{x<1}, \\
(d) \quad & \frac{d(\coth^{-1} x)}{dx} = \frac{1}{1-x^2}\bigg]_{x>1}, \\
(e) \quad & \frac{d(\operatorname{sech}^{-1} x)}{dx} = -\frac{1}{x\sqrt{1-x^2}}, \\
(f) \quad & \frac{d(\operatorname{csch}^{-1} x)}{dx} = -\frac{1}{x\sqrt{x^2+1}},
\end{aligned} \qquad (26)
$$

(a) Let $u = \sinh^{-1} x$, then $x = \sinh u$, $dx = \cosh u \, du = \sqrt{1+\sinh^2 u} = \sqrt{1+x^2} \, du$, $du = \dfrac{dx}{\sqrt{1+x^2}}$.

(b) Similar to (a).

(c) Let $u = \tanh^{-1} x$, then $x = \tanh u$, $dx = \operatorname{sech}^2 u \, du = (1 - \tanh^2 u) du = (1-x^2) du$, $du = \dfrac{dx}{1-x^2}$.

(d) Similar to (c).

(e)
$$\frac{d(\operatorname{sech}^{-1} x)}{dx} = \frac{d}{dx}\left(\cosh^{-1} \frac{1}{x}\right) = \frac{\frac{-1}{x^2}}{\left(\frac{1}{x^2} - 1\right)^{\frac{1}{2}}} = \frac{-1}{x\sqrt{1-x^2}}.$$

ARTICLE 15. DERIVATIVES OF ANTI-HYPERBOLIC FUNCTIONS.

(f) Similar to (e).

Prob. 31. Prove

$$\frac{d(\sin^{-1} x)}{dx} = \frac{1}{\sqrt{1-x^2}}, \qquad \frac{d(\cos^{-1} x)}{dx} = -\frac{1}{\sqrt{1-x^2}},$$

$$\frac{d(\tan^{-1} x)}{dx} = \frac{1}{1+x^2}, \qquad \frac{d(\cot^{-1} x)}{dx} = -\frac{1}{1+x^2}.$$

Prob. 32. Prove

$$d\sinh^{-1}\frac{x}{a} = \frac{dx}{\sqrt{x^2+a^2}}, \qquad d\cosh^{-1}\frac{x}{a} = \frac{dx}{\sqrt{x^2-a^2}},$$

$$d\tanh^{-1}\frac{x}{a} = \frac{adx}{a^2-x^2}\bigg]_{x<a}, \qquad d\coth^{-1}\frac{x}{a} = -\frac{adx}{x^2-a^2}\bigg]_{x>a}.$$

Prob. 33. Find $d(\text{sech}^{-1} x)$ independently of $\cosh^{-1} x$.

Prob. 34. When $\tanh^{-1} x$ is real, prove that $\coth^{-1} x$ is imaginary, and conversely; except when $x = 1$.

Prob. 35. Evaluate $\dfrac{\sinh^{-1} x}{\log x}$, $\dfrac{\cosh^{-1} x}{\log x}$ when $x = \infty$.

Article 16

Expansion of Hyperbolic Functions.

For this purpose take Maclaurin's Theorem,

$$f(u) = f(0) + uf'(0) + \frac{1}{2!}u^2 f''(0) + \frac{1}{3!}u^3 f'''(0) + \ldots,$$

and put

$$f(u) = \sinh u, \quad f'(u) = \cosh u, \quad f''(u) = \sinh u, \ldots,$$

then

$$f(0) = \sinh 0 = 0, \quad f'(0) = \cosh 0 = 1, \ldots;$$

hence

$$\sinh u = u + \frac{1}{3!}u^3 + \frac{1}{5!}u^5 + \ldots; \tag{27}$$

and similarly, or by differentiation,

$$\cosh u = 1 + \frac{1}{2!}u^2 + \frac{1}{4!}u^4 + \ldots. \tag{28}$$

By means of these series the numerical values of $\sinh u, \cosh u$, can be computed and tabulated for successive values of the independent variable u. They are convergent for all values of u, because the ratio of the nth term to the preceding is in the first case $\dfrac{u^2}{(2n-1)(2n-2)}$, and in the second case $\dfrac{u^2}{(2n-2)(2n-3)}$, both of which ratios can be made less than unity by taking n large enough, no

ARTICLE 16. EXPANSION OF HYPERBOLIC FUNCTIONS.

matter what value u has. Lagrange's remainder shows equivalence of function and series.

From these series the following can be obtained by division:

$$\left.\begin{aligned}
\tanh u &= u - \frac{1}{3}u^3 + \frac{2}{15}u^5 + \frac{17}{315}u^7 + \ldots, \\
\operatorname{sech} u &= 1 - \frac{1}{2}u^2 + \frac{5}{24}u^4 - \frac{61}{720}u^6 + \ldots, \\
u \coth u &= 1 + \frac{1}{3}u^2 - \frac{1}{45}u^4 + \frac{2}{945}u^6 - \ldots, \\
u \operatorname{csch} u &= 1 - \frac{1}{6}u^2 + \frac{7}{360}u^4 - \frac{31}{15120}u^6 + \ldots.
\end{aligned}\right\} \quad (29)$$

These four developments are seldom used, as there is no observable law in the coefficients, and as the functions $\tanh u, \operatorname{sech} u, \coth u, \operatorname{csch} u$, can be found directly from the previously computed values of $\cosh u, \sinh u$.

Prob. 36. Show that these six developments can be adapted to the circular functions by changing the alternate signs.

Article 17

Exponential Expressions.

Adding and subtracting (27), (28) give the identities

$$\cosh u + \sinh u = 1 + u + \frac{1}{2!}u^2 + \frac{1}{3!}u^3 + \frac{1}{4!}u^4 + \ldots = e^u,$$
$$\cosh u - \sinh u = 1 - u + \frac{1}{2!}u^2 - \frac{1}{3!}u^3 + \frac{1}{4!}u^4 - \ldots = e^{-u},$$

hence

$$\begin{rcases} \cosh u = \tfrac{1}{2}(e^u + e^{-u}), \quad \sinh u = \tfrac{1}{2}(e^u - e^{-u}), \\ \tanh u = \dfrac{e^u - e^{-u}}{e^u + e^{-u}}, \quad \operatorname{sech} u = \dfrac{2}{e^u + e^{-u}}, \quad \text{etc.} \end{rcases} \quad (30)$$

The analogous exponential expressions for $\sin u, \cos u$ are

$$\cos u = \frac{1}{2}(e^{ui} + e^{-ui}), \quad \sin u = \frac{1}{2i}(e^{u} - e^{-ui}), \quad (i = \sqrt{-1})$$

where the symbol e^{ui} stands for the result of substituting ui for x in the exponential development

$$e^x = 1 + x + \frac{1}{2!}x^2 + \frac{1}{3!}x^3 + \ldots$$

This will be more fully explained in treating of complex numbers, Arts. 28, 29.

Prob. 37. Show that the properties of the hyperbolic functions could be placed on a purely algebraic basis by starting with equations (30) as their definitions; for example, verify the identities:

$$\sinh(-u) = -\sinh u, \quad \cosh(-u) = \cosh u,$$
$$\cosh^2 u - \sinh^2 u = 1, \quad \sinh(u+v) = \sinh u \cosh v + \cosh u \sinh v,$$
$$\frac{d^2(\cosh mu)}{du^2} = m^2 \cosh mu, \quad \frac{d^2(\sinh mu)}{du^2} = m^2 \sinh mu.$$

ARTICLE 17. EXPONENTIAL EXPRESSIONS.

Prob. 38. Prove $(\cosh u + \sinh u)^n = \cosh nu + \sinh nu$.

Prob. 39. Assuming from Art. 14 that $\cosh u$, $\sinh u$ satisfy the differential equation $\dfrac{d^2y}{du^2} = y$, whose general solution may be written $y = Ae^n + Be^{-n}$, where A, B are arbitrary constants; show how to determine A, B in order to derive the expressions for $\cosh u$, $\sinh u$, respectively. [Use eq. (15).]

Prob. 40. Show how to construct a table of exponential functions from a table of hyperbolic sines and cosines, and *vice versa*.

Prob. 41. Prove $u = \log_e(\cosh u + \sinh u)$.

Prob. 42. Show that the area of any hyperbolic sector is infinite when its terminal line is one of the asymptotes.

Prob. 43. From the relation $2\cosh u = e^n + e^{-n}$ prove

$$2^{n-1}(\cosh u)^n = \cosh nu + n\cosh(n-2)u + \tfrac{1}{2}n(n-1)\cosh(n-4)u + \ldots,$$

and examine the last term when n is odd or even. Find also the corresponding expression for $2^{n-1}(\sinh u)^n$.

Article 18

Expansion of Anti-functions.

Since
$$\frac{d(\sinh^{-1} x)}{dx} = \frac{1}{\sqrt{1+x^2}} = (1+x^2)^{-\frac{1}{2}}$$
$$= 1 - \frac{1}{2} \cdot x^2 + \frac{1}{2} \cdot \frac{3}{4} \cdot x^4 - \frac{1}{2} \cdot \frac{3}{4} \cdot \frac{5}{6} \cdot x^6 + \ldots,$$

hence, by integration,

$$\sinh^{-1} x = x - \frac{1}{2} \cdot \frac{x^3}{3} + \frac{1}{2} \cdot \frac{3}{4} \cdot \frac{x^5}{5} - \frac{1}{2} \cdot \frac{3}{4} \cdot \frac{5}{6} \cdot \frac{x^7}{7} + \ldots, \qquad (31)$$

the integration-constant being zero, since $\sinh^{-1} x$ vanishes with x. This series is convergent, and can be used in computation, only when $x < 1$. Another series, convergent when $x > 1$, is obtained by writing the above derivative in the form

$$\frac{d(\sinh^{-1} x)}{dx} = (x^2+1)^{-\frac{1}{2}} = \frac{1}{x}\left(1 + \frac{1}{x^2}\right)^{-\frac{1}{2}}$$
$$= \frac{1}{x}\left[1 - \frac{1}{2} \cdot \frac{1}{x^2} + \frac{1}{2} \cdot \frac{3}{4} \cdot \frac{1}{x^4} - \frac{1}{2} \cdot \frac{3}{4} \cdot \frac{5}{6} \cdot \frac{1}{x^6} + \ldots\right],$$
$$\therefore \sinh^{-1} = C + \log x + \frac{1}{2} \cdot \frac{1}{2x^2} - \frac{1}{2} \cdot \frac{3}{4} \cdot \frac{1}{4x^4} + \frac{1}{2} \cdot \frac{3}{4} \cdot \frac{5}{6} \cdot \frac{1}{6x^6} - \ldots, \qquad (32)$$

where C is the integration-constant, which will be shown in Art. 19 to be equal to $\log_e 2$.

ARTICLE 18. EXPANSION OF ANTI-FUNCTIONS.

A development of similar form is obtained for $\cosh^{-1} x$; for

$$\frac{d(\cosh^{-1} x)}{dx} = (x^2 - 1)^{-\frac{1}{2}} = \frac{1}{x}\left(1 - \frac{1}{x^2}\right)^{-\frac{1}{2}}$$

$$= \frac{1}{x}\left[1 + \frac{1}{2} \cdot \frac{1}{x^2} + \frac{1}{2} \cdot \frac{3}{4} \cdot \frac{1}{x^4} + \frac{1}{2} \cdot \frac{3}{4} \cdot \frac{5}{6} \cdot \frac{1}{x^6} + \cdots\right],$$

hence

$$\cosh^{-1} x = C + \log x - \frac{1}{2} \cdot \frac{1}{2x^2} - \frac{1}{2} \cdot \frac{3}{4} \cdot \frac{1}{4x^4} - \frac{1}{2} \cdot \frac{3}{4} \cdot \frac{5}{6} \cdot \frac{1}{6x^6} - \cdots, \quad (33)$$

in which C is again equal to $\log_e 2$ [Art. 19, Prob. 46]. In order that the function $\cosh^{-1} x$ may be real, x must not be less than unity; but when x exceeds unity, this series is convergent, hence it is always available for computation.

Again

$$\frac{d(\tanh^{-1} x)}{dx} = \frac{1}{1 - x^2} = 1 + x^2 + x^4 + x^6 + \cdots,$$

and hence

$$\tanh^{-1} x = x + \frac{1}{3}x^3 + \frac{1}{5}x^5 + \frac{1}{7}x^7 + \cdots, \quad (34)$$

From (32), (33), (34) are derived:

$$\operatorname{sech}^{-1} x = \cosh^{-1} \frac{1}{x}$$

$$= C - \log x - \frac{x^2}{2 \cdot 2} - \frac{1 \cdot 3 \cdot x^4}{2 \cdot 4 \cdot 4} - \frac{1 \cdot 3 \cdot 5 \cdot x^6}{2 \cdot 4 \cdot 6 \cdot 6} - \cdots; \quad (35)$$

$$\operatorname{csch}^{-1} x = \sinh^{-1} \frac{1}{x} = \frac{1}{x} - \frac{1}{2} \cdot \frac{1}{3x^3} + \frac{1}{2} \cdot \frac{3}{4} \cdot \frac{1}{5x^5} - \frac{1}{2} \cdot \frac{3}{4} \cdot \frac{5}{6} \cdot \frac{1}{7x^7} + \cdots,$$

$$= C - \log x + \frac{x^2}{2 \cdot 2} - \frac{1 \cdot 3 \cdot x^4}{2 \cdot 4 \cdot 4} + \frac{1 \cdot 3 \cdot 5 \cdot x^6}{2 \cdot 4 \cdot 6 \cdot 6} - \cdots; \quad (36)$$

$$\coth^{-1} x = \tanh^{-1} \frac{1}{x} = \frac{1}{x} + \frac{1}{3x^3} + \frac{1}{5x^5} + \frac{1}{7x^7} + \cdots. \quad (37)$$

Prob. 44. Show that the series for $\tanh^{-1} x, \coth^{-1} x, \operatorname{sech}^{-1} x$, are always available for computation.

Prob. 45. Show that one or other of the two developments of the inverse hyperbolic cosecant is available.

Article 19

Logarithmic Expression of Anti-Functions.

Let
$$x = \cosh u,$$
then
$$\sqrt{x^2 - 1} = \sinh u;$$
therefore
$$x + \sqrt{x^2 - 1} = \cosh u + \sinh u = e^u,$$
and
$$u = \cosh^{-1} x = \log\left(x + \sqrt{x^2 - 1}\right). \tag{38}$$

Similarly,
$$\sinh^{-1} x = \log\left(x + \sqrt{x^2 + 1}\right). \tag{39}$$

Also
$$\operatorname{sech}^{-1} x = \cosh^{-1} \frac{1}{x} = \log \frac{1 + \sqrt{1 - x^2}}{x}, \tag{40}$$
$$\operatorname{csch}^{-1} x = \sinh^{-1} \frac{1}{x} = \log \frac{1 + \sqrt{1 + x^2}}{x}. \tag{41}$$

ARTICLE 19. LOGARITHMIC EXPRESSION OF ANTI-FUNCTIONS.

Again, let

$$x = \tanh u = \frac{e^u - e^{-u}}{e^u + e^{-u}},$$

therefore

$$\frac{1+x}{1-x} = \frac{e^u}{e^{-u}} = e^{2u},$$

$$2u = \log \frac{1+x}{1-x}, \quad \tanh^{-1} = \tfrac{1}{2} \log \frac{1+x}{1-x}; \tag{42}$$

and

$$\coth^{-1} x = \tanh^{-1} \frac{1}{x} = \tfrac{1}{2} \log \frac{x+1}{x-1}. \tag{43}$$

Prob. 46. Show from (38), (39), that, when $x \doteq \infty$,

$$\sinh^{-1} x - \log x \doteq \log 2, \quad \cosh^{-1} x - \log x \doteq \log 2,$$

and hence show that the integration-constants in (32), (33) are each equal to $\log 2$.

Prob. 47. Derive from (42) the series for $\tanh^{-1} x$ given in (34).

Prob. 48. Prove the identities:

$$\log x = 2\tanh^{-1} \frac{x-1}{x+1} = \tanh^{-1} \frac{x^2-1}{x^2+1} = \sinh^{-1} \tfrac{1}{2}(x - x^{-1}) = \cosh^{-1} \tfrac{1}{2}(x + x^{-1});$$

$$\log \sec x = 2\tanh^{-1} \tfrac{1}{2}x; \quad \log \csc x = 2\tanh^{-1} \tan^2 \left(\tfrac{1}{4}\pi + \tfrac{1}{2}x\right);$$

$$\log \tan x = -\tanh^{-1} \cos 2x = -\sinh^{-1} \cot 2x = \cosh^{-1} \csc 2x.$$

Article 20

The Gudermanian Function.

The correspondence of sectors of the same species was discussed in Arts. 1–4. It is now convenient to treat of the correspondence that may exist between sectors of different species.

Two points P_1, P_2, on any hyperbola and ellipse, are said to correspond with reference to two pairs of conjugates O_1A_1, O_1B_1, and O_2A_2, O_2B_2, respectively, when

$$\frac{x_1}{a_1} = \frac{a_2}{x_2}, \qquad (44)$$

and when y_1, y_2 have the same sign. The sectors $A_1O_1P_1, A_2O_2P_2$ are then also said to correspond. Thus corresponding sectors of central conics of different species are of the same sign and have their primary characteristic ratios reciprocal. Hence there is a fixed functional relation between their respective measures. The elliptic sectorial measure is called the gudermanian of the corresponding hyperbolic sectorial measure, and the latter the anti-gudermanian of the former. This relation is expressed by

$$\frac{S_2}{K_2} = \operatorname{gd} \frac{S_1}{K_1}$$

or $v = \operatorname{gd} u$, and $u = \operatorname{gd}^{-1} v$. $\qquad (45)$

Article 21

Circular Functions of Gudermanian.

The six hyperbolic functions of u are expressible in terms of the six circular functions of its gudermanian; for since

$$\frac{x_1}{a_1} = \cosh u, \quad \frac{x_2}{a_2} = \cosh v, \qquad \text{(see Arts. 6, 7)}$$

in which u, v are the measures of corresponding hyperbolic and elliptic sectors, hence

$$\left.\begin{aligned} \cosh u &= \sec v, \quad [\text{eq. (44)}] \\ \sinh u &= \sqrt{\sec^2 v - 1} = \tan v, \\ \tanh u &= \frac{\tan v}{\sec v} = \sin v, \\ \coth u &= \csc v, \\ \operatorname{sech} u &= \cos v, \\ \operatorname{csch} u &= \cot v. \end{aligned}\right\} \qquad (46)$$

The gudermanian is sometimes useful in computation; for instance, if $\sinh u$ be given, v can be found from a table of natural tangents, and the other circular functions of v will give the remaining hyperbolic functions of u. Other uses of this function are given in Arts. 22–26, 32–36.

Prob. 49. Prove that

$$\operatorname{gd} u = \sec^{-1}(\cosh u) = \tan^{-1}(\sinh u)$$
$$= \cos^{-1}(\operatorname{sech} u) = \sin^{-1}(\tanh u).$$

Prob. 50. Prove

$$\operatorname{gd}^{-1} v = \cosh^{-1}(\sec v) = \sinh^{-1}(\tan v)$$
$$= \operatorname{sech}^{-1}(\cos v) = \tanh^{-1}(\sin v).$$

ARTICLE 21. CIRCULAR FUNCTIONS OF GUDERMANIAN.

Prob. 51. Prove
$$\text{gd}\, 0 = 0, \ \text{gd}\, \infty = \tfrac{1}{2}\pi, \ \text{gd}(-\infty) = -\tfrac{1}{2}\pi,$$
$$\text{gd}^{-1} 0 = 0, \ \text{gd}^{-1}\left(\tfrac{1}{2}\pi\right) = \infty, \ \text{gd}^{-1}\left(-\tfrac{1}{2}\pi\right) = -\infty.$$

Prob. 52. Show that $\text{gd}\, u$ and $\text{gd}^{-1} v$ are odd functions of u, v.

Prob. 53. From the first identity in 4, Prob. 17, derive the relation $\tanh \tfrac{1}{2}u = \tan \tfrac{1}{2}v$.

Prob. 54. Prove $\tanh^{-1}(\tan u) = \tfrac{1}{2}\text{gd}\, 2u$, and $\tan^{-1}(\tanh x) = \tfrac{1}{2}\text{gd}^{-1} 2x$.

Article 22

Gudermanian Angle

If a circle be used instead of the ellipse of Art. 20, the gudermanian of the hyperbolic sectorial measure will be equal to the radian measure of the angle of the corresponding circular sector (see eq. (6), and Art. 3, Prob. 2). This angle will be called the gudermanian angle; but the gudermanian function v, as above defined, is merely a number, or ratio; and this number is equal to the radian measure of the gudermanian angle θ, which is itself usually tabulated in degree measure; thus

$$\theta = \frac{180° v}{\pi} \quad (47)$$

Prob. 55. Show that the gudermanian angle of u may be constructed as follows:

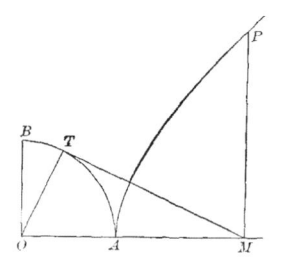

Take the principal radius OA of an equilateral hyperbola, as the initial line, and OP as the terminal line, of the sector whose measure is u; from M, the foot of the ordinate of P, draw MT tangent to the circle whose diameter is the transverse axis; then AOT is the angle required.[1]

Prob. 56. Show that the angle θ never exceeds 90°.

---

[1]This angle was called by Gudermann the longitude of u, and denoted by lu. His inverse symbol was $\mathfrak{L}$; thus $u = \mathfrak{L}(lu)$. (Crelle's Journal, vol. 6, 1830.) Lambert, who introduced the angle θ, named it the transcendent angle. (Hist. de l'acad. roy. de Berlin, 1761). Hoüel (Nouvelles Annales, vol. 3, 1864) called it the hyperbolic amplitude of u, and wrote it amh u, in analogy with the amplitude of an elliptic function, as shown in Prob. 62. Cayley (Elliptic Functions, 1876) made the usage uniform by attaching to the angle the name of the mathematician who had used it extensively in tabulation and in the theory of elliptic functions of modulus unity.

Prob. 57. The bisector of angle AOT bisects the sector AOP (see Prob. 13, Art. 9, and Prob. 53, Art. 21), and the line AP. (See Prob. 1, Art. 3.)

Prob. 58. This bisector is parallel to TP, and the points T, P are in line with the point diametrically opposite to A.

Prob. 59. The tangent at P passes through the foot of the ordinate of T, and intersects TM on the tangent at A.

Prob. 60. The angle APM is half the gudermanian angle.

Article 23

Derivatives of Gudermanian and Inverse.

Let
$$v = \operatorname{gd} u, \quad u = \operatorname{gd}^{-1} v,$$
then
$$\sec v = \cosh u,$$
$$\sec v \tan v \, dv = \sinh u \, du,$$
$$\sec v \, dv = du,$$
therefore
$$d(\operatorname{gd}^{-1} v) = \sec v \, dv. \tag{48}$$

Again,
$$dv = \cos v \, du = \operatorname{sech} u \, du,$$
therefore
$$d(\operatorname{gd} u) = \operatorname{sech} u \, du. \tag{49}$$

Prob. 61. Differentiate:
$$y = \sinh u - \operatorname{gd} u, \qquad y = \sin v + \operatorname{gd}^{-1} v,$$
$$y = \tanh u \operatorname{sech} u + \operatorname{gd} u, \qquad y = \tan v \sec v + \operatorname{gd}^{-1} v.$$

ARTICLE 23. DERIVATIVES OF GUDERMANIAN AND INVERSE.

Prob. 62. Writing the "elliptic integral of the first kind" in the form

$$u = \int_0^\phi \frac{d\phi}{\sqrt{1 - \kappa^2 \sin^2 \phi}},$$

κ being called the modulus, and ϕ the amplitude; that is,

$$\phi = \operatorname{am} u, (\operatorname{mod.} \kappa),$$

show that, in the special case when $\kappa = 1$,

$$u = \operatorname{gd}^{-1} \phi, \qquad \operatorname{am} u = \operatorname{gd} u, \qquad \sin \operatorname{am} u = \tanh u,$$
$$\cos \operatorname{am} u = \operatorname{sech} u, \qquad \tan \operatorname{am} u = \sinh u;$$

and that thus the elliptic functions $\sin \operatorname{am} u$, etc., degenerate into the hyperbolic functions, when the modulus is unity.[1]

---

[1]The relation $\operatorname{gd} u = \operatorname{am} u, (\operatorname{mod.} 1)$, led Hoüel to name the function $\operatorname{gd} u$, the hyperbolic amplitude of u, and to write it $\operatorname{amh} u$ (see note, Art. 22). In this connection Cayley expressed the functions $\tanh u$, $\operatorname{sech} u$, $\sinh u$ in the form $\sin \operatorname{gd} u$, $\cos \operatorname{gd} u$, $\tan \operatorname{gd} u$, and wrote them $\operatorname{sg} u$, $\operatorname{cg} u$, $\operatorname{tg} u$, to correspond with the abbreviations $\operatorname{sn} u$, $\operatorname{cn} u$, $\operatorname{dn} u$ for $\sin \operatorname{am} u$, $\cos \operatorname{am} u$, $\tan \operatorname{am} u$. Thus $\tanh u = \operatorname{sg} u = \operatorname{sn} u, (\operatorname{mod.} 1)$; etc.

It is well to note that neither the elliptic nor the hyperbolic functions received their names on account of the relation existing between them in a special case. (See foot-note, p. 1)

Article 24

Series for Gudermanian and its Inverse.

Substitute for $\text{sech}\, u, \sec v$ in (49), (48) their expansions, Art. 16, and integrate, then

$$\text{gd}\, u = u - \frac{1}{6}u^3 + \frac{1}{24}u^5 - \frac{61}{5040}u^7 + \cdots \tag{50}$$

$$\text{gd}^{-1} v = v + \frac{1}{6}v^3 + \frac{1}{24}v^5 - \frac{61}{5040}v^7 + \cdots \tag{51}$$

No constants of integration appear, since $\text{gd}\, u$ vanishes with u, and $\text{gd}^{-1} v$ with v. These series are seldom used in computation, as $\text{gd}\, u$ is best found and tabulated by means of tables of natural tangents and hyperbolic sines, from the equation

$$\text{gd}\, u = \tan^{-1}(\sinh u),$$

and a table of the direct function can be used to furnish the numerical values of the inverse function; or the latter can be obtained from the equation,

$$\text{gd}^{-1} v = \sinh^{-1}(\tan v) = \cosh^{-1}(\sec v).$$

To obtain a logarithmic expression for $\text{gd}^{-1} v$, let

$$\text{gd}^{-1} v = u, \quad v = \text{gd}\, u,$$

therefore

$$\sec v = \cosh u, \quad \tan v = \sinh u,$$
$$\sec v + \tan v = \cosh u + \sinh u = e^u,$$
$$e^u = \frac{1 + \sin v}{\cos v} = \frac{1 - \cos(\frac{1}{2}\pi + v)}{\sin(\frac{1}{2}\pi + v)} = \tan\left(\frac{1}{4}\pi + \frac{1}{2}v\right),$$
$$u = \text{gd}^{-1} v = \log_e \tan\left(\frac{1}{4}\pi + \frac{1}{2}v\right). \tag{52}$$

ARTICLE 24. SERIES FOR GUDERMANIAN AND ITS INVERSE.

Prob. 63. Evaluate $\left.\dfrac{\operatorname{gd} u - u}{u^3}\right]_{u \doteq 0}$, $\left.\dfrac{\operatorname{gd}^{-1} v - v}{v^3}\right]_{v \doteq 0}$.

Prob. 64. Prove that $\operatorname{gd} u - \sin u$ is an infinitesimal of the fifth order, when $u \doteq 0$.

Prob. 65. Prove the relations $\tfrac{1}{4}\pi + \tfrac{1}{2}v \tan^{-1} e^u$, $\tfrac{1}{4}\pi - \tfrac{1}{2}v = \tan^{-1} e^{-u}$.

Article 25

Graphs of Hyperbolic Functions.

Drawing two rectangular axes, and laying down a series of points whose abscissas represent, on any convenient scale, successive values of the sectorial measure, and whose ordinates represent, preferably on the same scale, the corresponding values of the function to be plotted, the locus traced out by this series of points will be a graphical representation of the variation of the function as the sectorial measure varies. The equations of the curves in the ordinary cartesian notation are:

Fig.	Full Lines.	Dotted Lines.
A	$y = \cosh x$,	$y = \operatorname{sech} x$;
B	$y = \sinh x$,	$y = \operatorname{csch} x$;
C	$y = \tanh x$,	$y = \coth x$;
D	$y = \operatorname{gd} x$.	

Here x is written for the sectorial measure u, and y for the numerical value of $\cosh u$, etc. It is thus to be noted that the variables x, y are numbers, or ratios, and that the equation $y = \cosh x$ merely expresses that the relation between the numbers x and y is taken to be the same as the relation between a sectorial measure and its characteristic ratio. The numerical values of $\cosh u$, $\sinh u$, $\tanh u$ are given in the tables at the end of this chapter for values of u between 0 and 4. For greater values they may be computed from the developments of Art. 16.

ARTICLE 25. GRAPHS OF HYPERBOLIC FUNCTIONS.

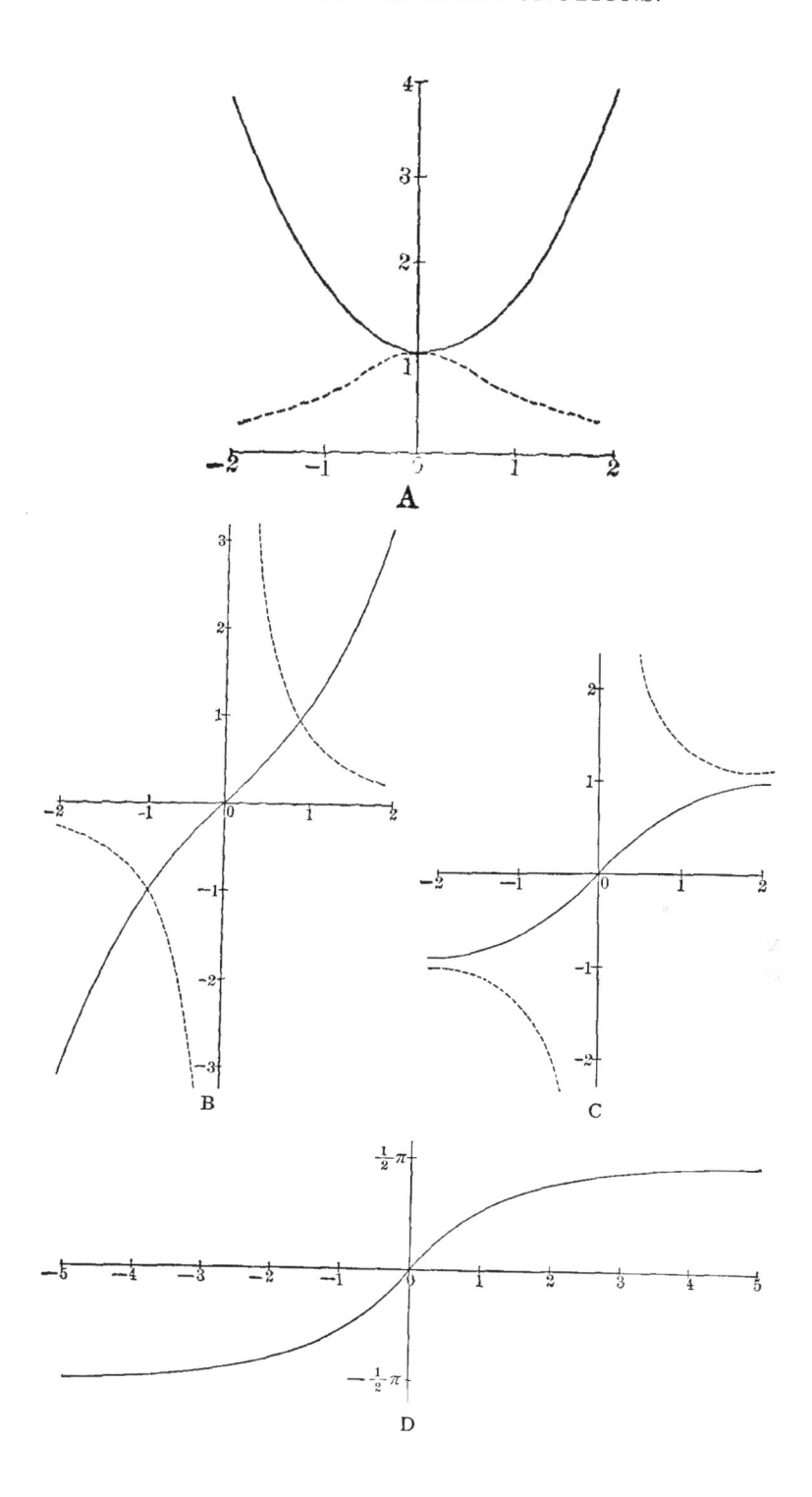

A

B

C

D

ARTICLE 25. GRAPHS OF HYPERBOLIC FUNCTIONS.

The curves exhibit graphically the relations:

$$\operatorname{sech} u = \frac{1}{\cosh u}, \quad \operatorname{csch} u = \frac{1}{\sinh u}, \quad \coth u = \frac{1}{\tanh u};$$

$\cosh u \not< 1, \quad \operatorname{sech} u \not> 1, \quad \tanh u \not> 1, \quad \operatorname{gd} u < \frac{1}{2}\pi, \text{ etc.};$

$\sinh(-u) = -\sinh u, \quad \cosh(-u) = \cosh u,$

$\tanh(-u) = -\tanh u, \quad \operatorname{gd}(-u) = -\operatorname{gd} u, \text{ etc.};$

$\cosh 0 = 1, \quad \sinh 0 = 0, \quad \tanh 0 = 0, \quad \operatorname{csch}(0) = \infty, \text{ etc.};$

$\cosh(\pm\infty) = \infty, \quad \sinh(\pm\infty) = \pm\infty, \quad \tanh(\pm\infty) = \pm 1, \text{ etc.}$

The slope of the curve $y = \sinh x$ is given by the equation $\frac{dy}{dx} = \cosh x$, showing that it is always positive, and that the curve becomes more nearly vertical as x becomes infinite. Its direction of curvature is obtained from $\frac{d^2y}{dx^2} = \sinh x$, proving that the curve is concave downward when x is negative, and upward when x is positive. The point of inflexion is at the origin, and the inflexional tangent bisects the angle between the axes.

The direction of curvature of the locus $y = \operatorname{sech} x$ is given by $\frac{d^2y}{dx^2} = \operatorname{sech} x(2\tanh^2 x - 1)$, and thus the curve is concave downwards or upwards according as $2\tanh^2 x - 1$ is negative or positive. The inflexions occur at the points $x = \pm\tanh^{-1}.707, = \pm.881, y = .707$; and the slopes of the inflexional tangents are $\mp\frac{1}{2}$.

The curve $y = \operatorname{csch} x$ is asymptotic to both axes, but approaches the axis of x more rapidly than it approaches the axis of y, for when $x = 3$, y is only .1, but it is not till $y = 10$ that x is so small as .1. The curves $y = \operatorname{csch} x$, $y = \sinh x$ cross at the points $x = \pm.881, y = \pm 1$.

Prob. 66. Find the direction of curvature, the inflexional tangent, and the asymptotes of the curves $y = \operatorname{gd} x$, $y = \tanh x$.

Prob. 67. Show that there is no inflexion-point on the curves $y = \cosh x$, $y = \coth x$.

Prob. 68. Show that any line $y = mx + n$ meets the curve $y = \tanh x$ in either three real points or one. Hence prove that the equation $\tanh x = mx + n$ has either three real roots or one. From the figure give an approximate solution of the equation $\tanh x = x - 1$.

Prob. 69. Solve the equations: $\cosh x = x + 2$; $\sinh x = \frac{3}{2}x$; $\operatorname{gd} x = x - \frac{1}{2}\pi$.

Prob. 70. Show which of the graphs represent even functions, and which of them represent odd ones.

Article 26

Elementary Integrals.

The following useful indefinite integrals follow from Arts. 14, 15, 23:

	Hyperbolic.	Circular.
1.	$\int \sinh u \, du = \cosh u,$	$\int \sin u \, du = -\cos u,$
2.	$\int \cosh u \, du = \sinh u,$	$\int \cos u \, du = \sin u,$
3.	$\int \tanh u \, du = \log \cosh u,$	$\int \tan u \, du = -\log \cos u,$
4.	$\int \coth u \, du = \log \sinh u,$	$\int \cot u \, du = \log \sin u,$
5.	$\int \operatorname{csch} u \, du = \log \tanh \frac{u}{2},$ $= -\sinh^{-1}(\operatorname{csch} u),$	$\int \csc u \, du = \log \tan \frac{u}{2},$ $= -\cosh^{-1}(\csc u),$
6.	$\int \operatorname{sech} u \, du = \operatorname{gd} u,$	$\int \sec u \, du = \operatorname{gd}^{-1} u,$
7.	$\int \dfrac{dx}{\sqrt{x^2+a^2}} = \sinh^{-1}\dfrac{x}{a},$ [1]	$\int \dfrac{dx}{\sqrt{a^2-x^2}} = \sin^{-1}\dfrac{x}{a},$
8.	$\int \dfrac{dx}{\sqrt{x^2-a^2}} = \cosh^{-1}\dfrac{x}{a},$	$\int \dfrac{-dx}{\sqrt{a^2-x^2}} = \cos^{-1}\dfrac{x}{a},$
9.	$\int \dfrac{dx}{a^2-x^2}\bigg]_{x<a} = \dfrac{1}{a}\tanh^{-1}\dfrac{x}{a},$	$\int \dfrac{dx}{a^2+x^2} = \dfrac{1}{a}\tan^{-1}\dfrac{x}{a},$
10.	$\int \dfrac{-dx}{x^2-a^2}\bigg]_{x>a} = \dfrac{1}{a}\coth^{-1}\dfrac{x}{a},$	$\int \dfrac{-dx}{a^2+x^2} = \dfrac{1}{a}\cot^{-1}\dfrac{x}{a},$
11.	$\int \dfrac{-dx}{x\sqrt{a^2-x^2}} = \dfrac{1}{a}\operatorname{sech}^{-1}\dfrac{x}{a},$	$\int \dfrac{dx}{x\sqrt{x^2-a^2}} = \dfrac{1}{a}\sec^{-1}\dfrac{x}{a},$
12.	$\int \dfrac{-dx}{x\sqrt{a^2+x^2}} = \dfrac{1}{a}\operatorname{csch}^{-1}\dfrac{x}{a},$	$\int \dfrac{-dx}{x\sqrt{x^2-a^2}} = \dfrac{1}{a}\csc^{-1}\dfrac{x}{a}.$

---

[1] Forms 7–12 are preferable to the respective logarithmic expressions (Art. 19), on account of the close analogy with the circular forms, and also because they involve functions that are directly tabulated. This advantage appears more clearly in 13–20.

ARTICLE 26. ELEMENTARY INTEGRALS.

From these fundamental integrals the following may be derived:

13. $\displaystyle\int \frac{dx}{\sqrt{ax^2+2bx+c}} = \frac{1}{\sqrt{a}}\sinh^{-1}\frac{ax+b}{\sqrt{ac-b^2}}$, a positive, $ac > b^2$;

$\displaystyle\qquad\qquad\qquad = \frac{1}{\sqrt{a}}\cosh^{-1}\frac{ax+b}{\sqrt{b^2-ac}}$, a positive, $ac < b^2$;

$\displaystyle\qquad\qquad\qquad = \frac{1}{\sqrt{-a}}\cos^{-1}\frac{ax+b}{\sqrt{b^2-ac}}$, a negative.

14. $\displaystyle\int \frac{dx}{ax^2+2bx+c} = \frac{1}{\sqrt{ac-b^2}}\tan^{-1}\frac{ax+b}{\sqrt{ac-b^2}}$, $ac > b^2$;

$\displaystyle\qquad\qquad\qquad = \frac{-1}{\sqrt{b^2-ac}}\tanh^{-1}\frac{ax+b}{\sqrt{b^2-ac}}$, $ac < b^2, ax+b < \sqrt{b^2-ac}$;

$\displaystyle\qquad\qquad\qquad = \frac{-1}{\sqrt{b^2-ac}}\coth^{-1}\frac{ax+b}{\sqrt{b^2-ac}}$, $ac < b^2, ax+b > \sqrt{b^2-ac}$;

Thus,

$$\int_4^5 \frac{dx}{x^2-4x+3} = -\coth^{-1}(x-2)\Big]_4^5 = \coth^{-1}2 - \coth^{-1}3$$
$$= \tanh^{-1}(.5) - \tanh^{-1}(.3333) = .5494 - .3466 = .2028.[2]$$

$$\int_2^{2.5} \frac{dx}{x^2-4x+3} = -\tanh^{-1}(x-2)\Big]_2^{2.5} = \tanh^{-1}0 - \tanh^{-1}(0.5) = -.5494.$$

(By interpreting these two integrals as areas, show graphically that the first is positive, and the second negative.)

15. $\displaystyle\int \frac{dx}{(a-x)\sqrt{x-b}} = \frac{2}{\sqrt{a-b}}\tanh^{-1}\sqrt{\frac{x-b}{a-b}}$,

$\displaystyle\qquad\qquad\qquad \text{or } \frac{-2}{\sqrt{b-a}}\tan^{-1}\sqrt{\frac{x-b}{b-a}}$,

$\displaystyle\qquad\qquad\qquad \text{or } \frac{2}{\sqrt{a-b}}\coth^{-1}\sqrt{\frac{x-b}{a-b}}$;

the real form to be taken. (Put $x - b = z^2$, and apply 9, 10.)

---

[2] For $\tanh^{-1}(.5)$ interpolate between $\tanh(.54) = .4930$, $\tanh(.56) = .5080$ (see tables, pp. 81, 82); and similarly for $\tanh^{-1}(.3333)$.

ARTICLE 26. ELEMENTARY INTEGRALS.

16. $\int \dfrac{dx}{(a-x)\sqrt{b-x}} = \dfrac{2}{\sqrt{b-a}} \tanh^{-1} \sqrt{\dfrac{b-x}{b-a}}$,

or $\dfrac{2}{\sqrt{b-a}} \coth^{-1} \sqrt{\dfrac{b-x}{b-a}}$,

or $\dfrac{-2}{\sqrt{a-b}} \tan^{-1} \sqrt{\dfrac{b-x}{a-b}}$;

the real form to be taken.

17. $\int (x^2 - a^2)^{\frac{1}{2}} dx = \dfrac{1}{2} x (x^2 - a^2)^{\frac{1}{2}} - \dfrac{1}{2} a^2 \cosh^{-1} \dfrac{x}{a}$.

By means of a reduction-formula this integral is easily made to depend on 8. It may also be obtained by transforming the expression into hyperbolic functions by the assumption $x = a \cosh u$, when the integral takes the form

$$a^2 \int \sinh^2 u \, du = \dfrac{a^2}{2} \int (\cosh 2u - 1) du = \dfrac{1}{4} a^2 (\sinh 2u - 2u)$$
$$= \dfrac{1}{2} a^2 (\sinh u \cosh u - u),$$

which gives 17 on replacing $a \cosh u$ by x, and $a \sinh u$ by $(x^2 - a^2)^{\frac{1}{2}}$. The geometrical interpretation of the result is evident, as it expresses that the area of a rectangular-hyperbolic segment AMP is the difference between a triangle OMP and a sector OAP.

18. $\int (a^2 - x^2)^{\frac{1}{2}} dx = \dfrac{1}{2} x (a^2 - x^2)^{\frac{1}{2}} + \dfrac{1}{2} a^2 \sin^{-1} \dfrac{x}{a}$.

19. $\int (x^2 + a^2)^{\frac{1}{2}} dx = \dfrac{1}{2} x (x^2 - a^2)^{\frac{1}{2}} + \dfrac{1}{2} a^2 \sinh^{-1} \dfrac{x}{a}$.

20. $\int \sec^3 \phi \, d\phi = \int (1 + \tan^2 \phi)^{\frac{1}{2}} d \tan \phi$

$$= \dfrac{1}{2} \tan \phi (1 + \tan^2 \phi)^{\frac{1}{2}} + \dfrac{1}{2} \sinh^{-1} (\tan \phi)$$
$$= \dfrac{1}{2} \sec \phi \tan \phi + \dfrac{1}{2} \mathrm{gd}^{-1} \phi.$$

21. $\int \mathrm{sech}^3 u \, du = \dfrac{1}{2} \mathrm{sech}\, u \tanh u + \dfrac{1}{2} \mathrm{gd}\, u$.

Prob. 71. What is the geometrical interpretation of 18, 19?

Prob. 72. Show that $\int (ax^2 + 2bx + c)^{\frac{1}{2}} dx$ reduces to 17, 18, 19, respectively: when a is positive, with $ac < b^2$; when a is negative; and when a is positive, with $ac > b^2$.

ARTICLE 26. ELEMENTARY INTEGRALS.

Prob. 73. Prove

$$\int \sinh u \tanh u \, du = \sinh u - \operatorname{gd} u,$$

$$\int \cosh u \coth u \, du = \cosh u + \log \tanh \frac{u}{2}.$$

Prob. 74. Integrate $(x^2 + 2x + 5)^{-\frac{1}{2}} dx$, $(x^2 + 2x + 5)^{-1} dx$, $(x^2 + 2x + 5)^{\frac{1}{2}} dx$.

Prob. 75. In the parabola $y^2 = 4px$, if s be the length of arc measured from the vertex, and ϕ the angle which the tangent line makes with the vertical tangent, prove that the intrinsic equation of the curve is $\dfrac{ds}{d\phi} = 2p \sec^3 \phi$, $s = p \sec \phi \tan \phi + p \operatorname{gd}^{-1} \phi$.

Prob. 76. The polar equation of a parabola being $r = a \sec^2 \theta$, referred to its focus as pole, express s in terms of θ.

Prob. 77. Find the intrinsic equation of the curve $\dfrac{y}{a} = \cosh \dfrac{x}{a}$, and of the curve $\dfrac{y}{a} = \log \sec \dfrac{x}{a}$.

Prob. 78. Investigate a formula of reduction for $\int \cosh^n x \, dx$; also integrate by parts $\cosh^{-1} x \, dx$, $\tanh^{-1} x \, dx$, $(\sinh^{-1} x)^2 dx$; and show that the ordinary methods of reduction for $\int \cos^m x \sin^n x \, dx$ can be applied to $\int \cosh^m x \sinh^n x \, dx$.

Article 27

Functions of Complex Numbers.

As vector quantities are of frequent occurrence in Mathematical Physics; and as the numerical measure of a vector in terms of a standard vector is a complex number of the form $x + iy$, in which x, y are real, and i stands for $\sqrt{-1}$; it becomes necessary in treating of any class of functional operations to consider the meaning of these operations when performed on such generalized numbers.[1] The geometrical definitions of $\cosh u$, $\sinh u$, given in Art. 7, being then no longer applicable, it is necessary to assign to each of the symbols $\cosh(x + iy)$, $\sinh(x + iy)$, a suitable algebraic meaning, which should be consistent with the known algebraic values of $\cosh x$, $\sinh x$, and include these values as a particular case when $y = 0$. The meanings assigned should also, if possible, be such as to permit the addition-formulas of Art. 11 to be made general, with all the consequences that flow from them.

Such definitions are furnished by the algebraic developments in Art. 16, which are convergent for all values of u, real or complex. Thus the definitions of $\cosh(x + iy)$, $\sinh(x + iy)$ are to be

$$\left.\begin{aligned}\cosh(x + iy) &= 1 + \frac{1}{2!}(x + iy)^2 + \frac{1}{4!}(x + iy)^4 + \ldots, \\ \sinh(x + iy) &= (x + iy) + \frac{1}{3!}(x + iy)^3 + \ldots\end{aligned}\right\} \quad (52)$$

From these series the numerical values of $\cosh(x + iy)$, $\sinh(x + iy)$ could be computed to any degree of approximation, when x and y are given. In general

---

[1] The use of vectors in electrical theory is shown in Bedell and Crehore's Alternating Currents, Chaps, XIV–XX (first published in 1892). The advantage of introducing the complex measures of such vectors into the differential equations is shown by Steinmetz, Proc. Elec. Congress, 1893; while the additional convenience of expressing the solution in hyperbolic functions of these complex numbers is exemplified by Kennelly, Proc. American Institute Electrical Engineers, April 1895. (See below, Art. 37.)

ARTICLE 27. FUNCTIONS OF COMPLEX NUMBERS.

the results will come out in the complex form[2]

$$\cosh(x+iy) = a+ib,$$
$$\sinh(x+iy) = c+id.$$

The other functions are defined as in Art. 7, eq. (9).

Prob. 79. Prove from these definitions that, whatever u may be,

$$\cosh(-u) = \cosh u, \qquad \sinh(-u) = -\sinh u,$$
$$\frac{d}{du}\cosh u = \sinh u, \qquad \frac{d}{du}\sinh u = \cosh u,$$
$$\frac{d^2}{du^2}\cosh mu = m^2 \cosh mu, \qquad \frac{d^2}{du^2}\sinh mu = m^2 \sinh mu.[3]$$

---

[2] It is to be borne in mind that the symbols cosh, sinh, here stand for algebraic operators which convert one number into another; or which, in the language of vector-analysis, change one vector into another, by stretching and turning.

[3] The generalized hyperbolic functions usually present themselves in Mathematical Physics as the solution of the differential equation $\frac{d^2\phi}{du^2} = m^2\phi$, where ϕ, m, u are complex numbers, the measures of vector quantities. (See Art. 37.)

Article 28

Addition-Theorems for Complexes.

The addition-theorems for $\cosh(u+v)$, etc., where u, v are complex numbers, may be derived as follows. First take u, v as real numbers, then, by Art. 11,

$$\cosh(u+v) = \cosh u \cosh v + \sinh u \sinh v;$$

hence

$$1 + \frac{1}{2!}(u+v)^2 + \ldots = \left(1 + \frac{1}{2!}u^2 + \ldots\right)\left(1 + \frac{1}{2!}v^2 + \ldots\right)$$
$$+ \left(u + \frac{1}{3!}u^3 + \ldots\right)\left(v + \frac{1}{3!}v^3 + \ldots\right)$$

This equation is true when u, v are any real numbers. It must, then, be an algebraic identity. For, compare the terms of the rth degree in the letters u, v on each side. Those on the left are $\frac{1}{r!}(u+v)^r$; and those on the right, when collected, form an rth-degree function which is numerically equal to the former for more than r values of u when v is constant, and for more than r values of v when u is constant. Hence the terms of the rth degree on each side are algebraically identical functions of u and v.[1] Similarly for the terms of any other degree. Thus the equation above written is an algebraic identity, and is true for all values of u, v, whether real or complex. Then writing for each side its symbol, it follows that

$$\cosh(u+v) = \cosh u \cosh v + \sinh u \sinh v; \tag{53}$$

and by changing v into $-v$,

$$\cosh(u-v) = \cosh u \cosh v - \sinh u \sinh v. \tag{54}$$

---

[1] "If two rth-degree functions of a single variable be equal for more than r values of the variable, then they are equal for all values of the variable, and are algebraically identical."

ARTICLE 28. ADDITION-THEOREMS FOR COMPLEXES.

In a similar manner is found

$$\sinh(u \pm v) = \sinh u \cosh v \pm \cosh u \sinh v. \tag{55}$$

In particular, for a complex argument,

$$\left.\begin{aligned}\cosh(x \pm iy) &= \cosh x \cosh iy \pm \sinh x \sinh iy,\\ \sinh(x \pm iy) &= \sinh x \cosh iy \pm \cosh x \sinh iy.\end{aligned}\right\} \tag{56}$$

Prob. 79. Show, by a similar process of generalization,[2] that if $\sin u$, $\cos u$, $\exp u$[3] be defined by their developments in powers of u, then, whatever u may be,

$$\sin(u + v) = \sin u \cos v + \cos u \sin v,$$
$$\cos(u + v) = \cos u \cos v - \sin u \sin v,$$
$$\exp(u + v) = \exp u \exp v.$$

Prob. 80. Prove that the following are identities:

$$\cosh^2 u - \sinh^2 u = 1,$$
$$\cosh u + \sinh u = \exp u,$$
$$\cosh u - \sinh u = \exp(-u),$$
$$\cosh u = \tfrac{1}{2}[\exp u + \exp(-u)],$$
$$\sinh u = \tfrac{1}{2}[\exp u - \exp(-u)].$$

---

[2] This method of generalization is sometimes called the principle of the "permanence of equivalence of forms." It is not, however, strictly speaking, a "principle," but a method; for, the validity of the generalization has to be demonstrated, for any particular form, by means of the principle of the algebraic identity of polynomials enunciated in the preceding foot-note. (See Annals of Mathematics, Vol. 6, p. 81.)

[3] The symbol $\exp u$ stands for "exponential function of u," which is identical with e^u when u is real.

Article 29

Functions of Pure Imaginaries.

In the defining identities

$$\cosh u = 1 + \frac{1}{2!}u^2 + \frac{1}{4!}u^4 + \cdots,$$
$$\sinh u = 1 + \frac{1}{3!}u^3 + \frac{1}{5!}u^5 + \cdots,$$

put for u the pure imaginary iy, then

$$\cosh iy = 1 - \frac{1}{2!}y^2 + \frac{1}{4!}y^4 - \cdots = \cos y, \tag{57}$$
$$\sinh iy = iy + \frac{1}{3!}(iy)^3 + \frac{1}{5!}(iy)^5 + \cdots$$
$$= i\left[y - \frac{1}{3!}y^3 + \frac{1}{5!}y^5 - \cdots\right] = i\sin y, \tag{58}$$

and, by division,

$$\tanh iy = i\tan y. \tag{59}$$

These formulas serve to interchange hyperbolic and circular functions. The hyperbolic cosine of a pure imaginary is real, and the hyperbolic sine and tangent are pure imaginaries.

The following table exhibits the variation of $\sinh u$, $\cosh u$. $\tanh u$, $\exp u$, as u takes a succession of pure imaginary values.

ARTICLE 29. FUNCTIONS OF PURE IMAGINARIES.

u	$\sinh u$	$\cosh u$	$\tanh u$	$\exp u$
0	0	1	0	1
$\frac{1}{4}i\pi$	$.7i$	$.7^1$	i	$.7(1+i)$
$\frac{1}{2}i\pi$	i	0	∞i	i
$\frac{3}{4}i\pi$	$.7i$	$-.7$	$-i$	$.7(1-i)$
$i\pi$	0	-1	0	-1
$\frac{5}{4}i\pi$	$-.7i$	$-.7$	i	$-.7(1+i)$
$\frac{3}{2}i\pi$	$-i$	0	∞i	$-i$
$\frac{7}{4}i\pi$	$-.7i$	$.7$	$-i$	$-.7(1-i)$
$2i\pi$	0	1	0	1

In this table .7 is written for $\frac{1}{2}\sqrt{2}, = .707\ldots$.

Prob. 81. Prove the following identities:

$$\cos y = \cosh iy = \frac{1}{2}\left[\exp iy + \exp(-iy)\right],$$
$$\sin y = \frac{1}{i}\sinh iy = \frac{1}{2i}\left[\exp iy - \exp(-iy)\right],$$
$$\cos y + i\sin y = \cosh iy + \sinh iy = \exp iy,$$
$$\cos y - i\sin y = \cosh iy - \sinh iy = \exp(-iy),$$
$$\cos iy = \cosh y, \quad \sin iy = i\sinh y.$$

Prob. 82. Equating the respective real and imaginary parts on each side of the equation $\cos ny + i\sin ny = (\cos y + i\sin y)^n$, express $\cos ny$ in powers of $\cos y$, $\sin y$; and hence derive the corresponding expression for $\cosh ny$.

Prob. 83. Show that, in the identities (57) and (58), y may be replaced by a general complex, and hence that

$$\sinh(x \pm iy) = \pm i\sin(y \mp ix),$$
$$\cosh(x \pm iy) = \cos(y \mp ix),$$
$$\sin(x \pm iy) = \pm i\sinh(y \mp ix),$$
$$\cos(x \pm iy) = \cosh(y \mp ix).$$

Prob. 84. From the product-series for $\sin x$ derive that for $\sinh x$:

$$\sin x = x\left(1 - \frac{x^2}{\pi^2}\right)\left(1 - \frac{x^2}{2^2\pi^2}\right)\left(1 - \frac{x^2}{3^2\pi^2}\right)\cdots,$$
$$\sinh x = x\left(1 + \frac{x^2}{\pi^2}\right)\left(1 + \frac{x^2}{2^2\pi^2}\right)\left(1 + \frac{x^2}{3^2\pi^2}\right)\cdots.$$

Article 30

Functions of $x + iy$ in the Form $X + iY$.

By the addition-formulas,

$$\cosh(x + iy) = \cosh x \cosh iy + \sinh x \sinh iy,$$
$$\sinh(x + iy) = \sinh x \cosh iy + \cosh x \sinh iy,$$

but

$$\cosh iy = \cos y, \quad \sinh iy = i \sin y,$$

hence

$$\left.\begin{array}{l} \cosh(x + iy) = \cosh x \cos y + i \sinh x \sin y, \\ \sinh(x + iy) = \sinh x \cos y + i \cosh x \sin y. \end{array}\right\} \qquad (60)$$

Thus if $\cosh(x + iy) = a + ib$, $\sinh(x + iy) = c + id$, then

$$\left.\begin{array}{ll} a = \cosh x \cos y, & b = \sinh x \sin y, \\ c = \sinh x \cos y, & d = \cosh x \sin y. \end{array}\right\} \qquad (61)$$

From these expressions the complex tables at the end of this chapter have been computed.

Writing $\cosh z = Z$, where $z = x + iy$, $Z = X + iY$; let the complex numbers z, Z be represented on Argand diagrams, in the usual way, by the points whose coordinates are (x, y), (X, Y); and let the point z move parallel to the y-axis, on a given line $x = m$, then the point Z will describe an ellipse whose equation, obtained by eliminating y between the equations $X = \cosh m \cos y$, $Y = \sinh m \sin y$, is

$$\frac{X^2}{(\cosh m)^2} + \frac{Y^2}{(\sinh m)^2} = 1,$$

and which, as the parameter m varies, represents a series of confocal ellipses, the distance between whose foci is unity. Similarly, if the point z move parallel to the x-axis, on a given line $y = n$, the point Z will describe an hyperbola whose equation, obtained by eliminating the variable x from the equations. $X = \cosh x \cos n$, $Y = \sinh x \sin n$, is

$$\frac{X^2}{(\cos n)^2} - \frac{Y^2}{(\sin n)^2} = 1,$$

and which, as the parameter n varies, represents a series of hyperbolas confocal with the former series of ellipses.

These two systems of curves, when accurately drawn at close intervals on the Z plane, constitute a chart of the hyperbolic cosine; and the numerical value of $\cosh(m + in)$ can be read off at the intersection of the ellipse whose parameter is m with the hyperbola whose parameter is n.[1] A similar chart can be drawn for $\sinh(x + iy)$, as indicated in Prob. 85.

Periodicity of Hyperbolic Functions.—The functions $\sinh u$ and $\cosh u$ have the pure imaginary period $2i\pi$. For

$$\sinh(u + 2i\pi) = \sinh u \cos 2\pi + i \cosh u \sin 2\pi = \sinh u,$$
$$\cosh(u + 2i\pi) = \cosh u \cos 2\pi + i \sinh u \sin 2\pi = \cosh u.$$

The functions $\sinh u$ and $\cosh u$ each change sign when the argument u is increased by the half period $i\pi$. For

$$\sinh(u + i\pi) = \sinh u \cos \pi + i \cosh u \sin \pi = -\sinh u,$$
$$\cosh(u + i\pi) = \cosh u \cos \pi + i \sinh u \sin \pi = -\cosh u.$$

The function $\tanh u$ has the period $i\pi$. For, it follows from the last two identities, by dividing member by member, that

$$\tanh(u + i\pi) = \tanh u.$$

By a similar use of the addition formulas it is shown that

$$\sinh(u + \frac{1}{2}i\pi) = i \cosh u, \quad \cosh(u + \frac{1}{2}i\pi) = i \sinh u.$$

By means of these periodic, half-periodic, and quarter-periodic relations, the hyperbolic functions of $x + iy$ are easily expressible in terms of functions of $x + iy'$, in which y' is less than $\frac{1}{2}\pi$.

---

[1] Such a chart is given by Kennelly, Proc. A. I. E. E., April 1895, and is used by him to obtain the numerical values of $\cosh(x + iy)$, $\sinh(x + iy)$, which present themselves as the measures of certain vector quantities in the theory of alternating currents. (See Art. 37.) The chart is constructed for values of x and of y between 0 and 1.2; but it is available for all values of y, on account of the periodicity of the functions.

ARTICLE 30. FUNCTIONS OF $X + IY$ IN THE FORM $X + IY$.

The hyperbolic functions are classed in the modern function-theory of a complex variable as functions that are singly periodic with a pure imaginary period, just as the circular functions are singly periodic with a real period, and the elliptic functions are doubly periodic with both a real and a pure imaginary period.

Multiple Values of Inverse Hyperbolic Functions.—It follows from the periodicity of the direct functions that the inverse functions $\sinh^{-1} m$ and $\cosh^{-1} m$ have each an indefinite number of values arranged in a series at intervals of $2i\pi$. That particular value of $\sinh^{-1} m$ which has the coefficient of i not greater than $\frac{1}{2}\pi$ nor less than $-\frac{1}{2}\pi$ is called the principal value of $\sinh^{-1} m$; and that particular value of $\cosh^{-1} m$ which has the coefficient of i not greater than π nor less than zero is called the principal value of $\cosh^{-1} m$. When it is necessary to distinguish between the general value and the principal value the symbol of the former will be capitalized; thus

$$\text{Sinh}^{-1} m = \sinh^{-1} m + 2ir\pi, \quad \text{Cosh}^{-1} m = \cosh^{-1} m + 2ir\pi,$$
$$\text{Tanh}^{-1} m = \tanh^{-1} m + ir\pi,$$

in which r is any integer, positive or negative.

Complex Roots of Cubic Equations.—It is well known that when the roots of a cubic equation are all real they are expressible in terms of circular functions. Analogous hyperbolic expressions are easily found when two of the roots are complex. Let the cubic, with second term removed, be written

$$x^3 \pm 3bx = 2c.$$

Consider first the case in which b has the positive sign. Let $x = r \sinh u$, substitute, and divide by r^3, then

$$\sinh^3 u + \frac{3b}{r^2} \sinh u = \frac{2c}{r^3}.$$

Comparison with the formula $\sinh^3 u + \frac{3}{4} \sinh u = \frac{1}{4} \sinh 3u$ gives

$$\frac{3b}{r^2} = \frac{3}{4}, \quad \frac{2c}{r^3} = \frac{\sinh 3u}{4},$$

whence

$$r = 2b^{\frac{1}{2}}, \quad \sinh 3u = \frac{c}{b^{\frac{3}{2}}}, \quad u = \frac{1}{3} \sinh^{-1} \frac{c}{b^{\frac{3}{2}}};$$

therefore

$$x = 2b^{\frac{1}{2}} \sinh \left(\frac{1}{3} \sinh^{-1} \frac{c}{b^{\frac{3}{2}}} \right),$$

in which the sign of $b^{\frac{1}{2}}$ is to be taken the same as the sign of c. Now let the principal value of $\sinh^{-1}\dfrac{c}{b^{\frac{3}{2}}}$, found from the tables, be n; then two of the imaginary values are $n \pm 2i\pi$, hence the three values of x are $2b^{\frac{1}{2}}\sinh\dfrac{n}{3}$ and $2b^{\frac{1}{2}}\sinh\left(\dfrac{n}{3}\pm\dfrac{2i\pi}{3}\right)$. The last two reduce to $-b^{\frac{1}{2}}\sinh\left(\dfrac{n}{3}\pm i\sqrt{3}\cosh\dfrac{n}{3}\right)$.

Next, let the coefficient of x be negative and equal to $-3b$. It may then be shown similarly that the substitution $x = r\sin\theta$ leads to the three solutions

$$-2b^{\frac{1}{2}}\sin\frac{n}{3},\quad b^{\frac{1}{2}}\left(\sin\frac{n}{3}\pm\sqrt{3}\cos\frac{n}{3}\right),\quad \text{where } n = \sin^{-1}\frac{c}{b^{\frac{3}{2}}}.$$

These roots are all real when $c \not> b^{\frac{3}{2}}$. If $c > b^{\frac{3}{2}}$, the substitution $x = r\cosh u$ leads to the solution

$$x = 2b^{\frac{1}{2}}\cosh\left(\frac{1}{3}\cosh^{-1}\frac{c}{b^{\frac{3}{2}}}\right),$$

which gives the three roots

$$2b^{\frac{1}{2}}\cosh\frac{n}{3},\quad -b^{\frac{1}{2}}\left(\cosh\frac{n}{3}\pm i\sqrt{3}\sinh\frac{n}{3}\right),\quad \text{wherein } n = \cosh^{-1}\frac{c}{b^{\frac{3}{2}}},$$

in which the sign of $b^{\frac{1}{2}}$ is to be taken the same as the sign of c.

Prob. 85. Show that the chart of $\cosh(x + iy)$ can be adapted to $\sinh(x + iy)$, by turning through a right angle; also to $\sin(x + iy)$.

Prob. 86. Prove the identity
$$\tanh(x+iy) = \frac{\sinh 2x + i\sin 2y}{\cosh 2x + \cos 2y}.$$

Prob. 87. If $\cosh(x + iy) = a + ib$, be written in the "modulus and amplitude" form as $r(\cos\theta + i\sin\theta), = r\exp i\theta$, then
$$r^2 = a^2 + b^2 = \cosh^2 x = \sin^2 y = \cos^2 y - \sinh^2 x,$$
$$\tan\theta = \frac{b}{a} = \tanh x\tan y.$$

Prob. 88. Find the modulus and amplitude of $\sinh(x + iy)$.

Prob. 89. Show that the period of $\exp\dfrac{2\pi u}{a}$ is ia.

Prob. 90. When m is real and > 1, $\cos^{-1} m = i\cosh^{-1} m$, $\sin^{-1} m = \frac{\pi}{2} - i\cosh^{-1} m$. When m is real and < 1, $\cosh^{-1} m = i\cos^{-1} m$.

Article 31

The Catenary

A flexible inextensible string is suspended from two fixed points, and takes up a position of equilibrium under the action of gravity. It is required to find the equation of the curve in which it hangs.

Let w be the weight of unit length, and s the length of arc AP measured from the lowest point A; then ws is the weight of the portion AP. This is balanced by the terminal tensions, T acting in the tangent line at P, and H in the horizontal tangent. Resolving horizontally and vertically gives

$$T\cos\phi = H, \quad T\sin\phi = ws,$$

in which ϕ is the inclination of the tangent at P; hence

$$\tan\phi = \frac{ws}{H} = \frac{s}{c},$$

where c is written for $\dfrac{H}{w}$, the length whose weight is the constant horizontal tension; therefore

$$\frac{dy}{dx} = \frac{s}{c}, \quad \frac{ds}{dx} = \sqrt{1+\frac{s^2}{c^2}}, \quad \frac{dx}{c} = \frac{ds}{\sqrt{s^2+c^2}},$$

$$\frac{x}{c} = \sinh^{-1}\frac{s}{c}, \quad \sinh\frac{x}{c} = \frac{s}{c} = \frac{dy}{dx}, \quad \frac{y}{c} = \cosh\frac{x}{c},$$

which is the required equation of the catenary, referred to an axis of x drawn at a distance c below A.

The following trigonometric method illustrates the use of the gudermanian: The "intrinsic equation," $s = c\tan\phi$, gives $ds = c\sec^2\phi\,d\phi$; hence $dx = ds\cos\phi = c\sec\phi\,d\phi$; $dy = ds\sin\phi = c\sec\phi\tan\phi\,d\phi$; thus $x = c\,\mathrm{gd}^{-1}\phi, y = c\sec\phi$; whence $\frac{y}{c} = \sec\phi = \sec\mathrm{gd}\frac{x}{c} = \cosh\frac{x}{c}$; and $\frac{s}{c} = \tan\mathrm{gd}\frac{x}{c} = \sinh\frac{x}{c}$.

Numerical Exercise.—A chain whose length is 30 feet is suspended from two points 20 feet apart in the same horizontal; find the parameter c, and the depth of the lowest point.

ARTICLE 31. THE CATENARY

The equation $\frac{s}{c} = \sinh \frac{x}{c}$ gives $\frac{15}{c} = \sinh \frac{10}{c}$, which, by putting $\frac{10}{c} = z$, may be written $1.5z = \sinh z$. By examining the intersection of the graphs of $y = \sinh z$, $y = 1.5z$, it appears that the root of this equation is $z = 1.6$, nearly. To find a closer approximation to the root, write the equation in the form $f(z) = \sinh z - 1.5z = 0$, then, by the tables,

$$f(1.60) = 2.3756 - 2.4000 = -.0244,$$
$$f(1.62) = 2.4276 - 2.4300 = -.0024,$$
$$f(1.64) = 2.4806 - 2.4600 = +.0206;$$

whence, by interpolation, it is found that $f(1.6221) = 0$, and $z = 1.6221$, $c = \frac{10}{z} = 6.1649$. The ordinate of either of the fixed points is given by the equation

$$\frac{y}{c} = \cosh \frac{x}{c} = \cosh \frac{10}{c} = \cosh 1.6221 = 2.6306,$$

from tables; hence $y = 16.2174$, and required depth of the vertex $= y - c = 10.0525$ feet.[1]

Prob. 91. In the above numerical problem, find the inclination of the terminal tangent to the horizon.

Prob. 92. If a perpendicular MN be drawn from the foot of the ordinate to the tangent at P, prove that MN is equal to the constant c, and that NP is equal to the arc AP. Hence show that the locus of N is the involute of the catenary, and has the property that the length of the tangent, from the point of contact to the axis of x, is constant. (This is the characteristic property of the tractory).

Prob. 93. The tension T at any point is equal to the weight of a portion of the string whose length is equal to the ordinate y of that point.

Prob. 94. An arch in the form of an inverted catenary[2] is 30 feet wide and 10 feet high; show that the length of the arch can be obtained from the equations $\cosh z - \frac{2}{3}z = i$, $2s = \frac{30}{z} \sinh z$.

---

[1] See a similar problem in Chap. 1, Art. 7.
[2] For the theory of this form of arch, see "Arch" in the Encyclopædia Britannica.

Article 32

Catenary of Uniform Strength.

If the area of the normal section at any point be made proportional to the tension at that point, there will then be a constant tension per unit of area, and the tendency to break will be the same at all points. To find the equation of the curve of equilibrium under gravity, consider the equilibrium of an element PP' whose length is ds, and whose weight is $g\rho\omega\,ds$, where ω is the section at P, and ρ the uniform density. This weight is balanced by the difference of the vertical components of the tensions at P and P', hence

$$d(T\sin\phi) = g\rho\omega\,ds,$$
$$d(T\cos\phi) = 0;$$

therefore $T\cos\phi = H$, the tension at the lowest point, and $T = H\sec\phi$. Again, if ω_0 be the section at the lowest point, then by hypothesis $\frac{\omega}{\omega_0} = \frac{T}{H} = \sec\phi$, and the first equation becomes

$$Hd(\sec\phi\sin\phi) = g\rho\omega_0\sec\phi\,ds,$$

or

$$c\,d\tan\phi = \sec\phi\,ds,$$

where c stands for the constant $\dfrac{H}{g\rho\omega_0}$, the length of string (of section ω_0) whose weight is equal to the tension at the lowest point; hence,

$$ds = c\sec\phi\,d\phi, \quad \frac{s}{c} = \text{gd}^{-1}\phi,$$

the intrinsic equation of the catenary of uniform strength.

Also

$$dx = ds\cos\phi = c\,d\phi, \quad dy = ds\sin\phi = c\tan\phi\,d\phi;$$

ARTICLE 32. CATENARY OF UNIFORM STRENGTH.

hence
$$x = c\phi, \quad y = c \log \sec \phi,$$

and thus the Cartesian equation is
$$\frac{y}{c} = \log \sec \frac{x}{c},$$

in which the axis of x is the tangent at the lowest point.

Prob. 95. Using the same data as in Art. 31, find the parameter c and the depth of the lowest point. (The equation $\frac{x}{c} = \operatorname{gd} \frac{s}{c}$ gives $\frac{10}{c} = \operatorname{gd} \frac{15}{c}$, which, by putting $\frac{15}{c} = z$, becomes $\operatorname{gd} z = \frac{2}{3}z$. From the graph it is seen that z is nearly 1.8. If $f(z) = \operatorname{gd} z - \frac{2}{3}z$, then, from the tables of the gudermanian at the end of this chapter,
$$f(1.80) = 1.2432 - 1.2000 = +.0432,$$
$$f(1.90) = 1.2739 - 1.2667 = +.0072,$$
$$f(1.95) = 1.2881 - 1.3000 = -.0119,$$

whence, by interpolation, $z = 1.9189$ and $c = 7.8170$. Again, $\frac{y}{c} = \log_e \sec \frac{x}{c}$; but $\frac{x}{c} = \frac{10}{c} = 1.2793$; and 1.2793 radians $= 73° \, 17' \, 55''$; hence $y = 7.8170 \times .41914 \times 2.3026 = 7.5443$, the required depth.)

Prob. 96. Find the inclination of the terminal tangent.

Prob. 97. Show that the curve has two vertical asymptotes.

Prob. 98. Prove that the law of the tension T, and of the section ω, at a distance s, measured from the lowest point along the curve, is
$$\frac{T}{H} = \frac{\omega}{\omega_0} = \cosh \frac{c}{h};$$
and show that in the above numerical example the terminal section is 3.48 times the minimum section.

Prob. 99. Prove that the radius of curvature is given by $\rho = c \cosh \frac{s}{f}$. Also that the weight of the arc s is given by $W = H \sinh \frac{s}{f}$, in which s is measured from the vertex.

Article 33

The Elastic Catenary.

An elastic string of uniform section and density in its natural state is suspended from two points. Find its equation of equilibrium.

Let the element $d\sigma$ stretch into ds; then, by Hooke's law, $ds = d\sigma(1 + \lambda T)$, where λ is the elastic constant of the string; hence the weight of the stretched element $ds = g\rho\omega\, d\sigma = \dfrac{g\rho\omega\, ds}{(1+\lambda T)}$. Accordingly, as before,

$$d(T\sin\phi) = \frac{g\rho\omega\, ds}{(1+\lambda T)},$$

and

$$T\cos\phi = H = g\rho\omega c,$$

hence

$$c\, d(\tan\phi) = \frac{ds}{(1+\mu\sec\phi)},$$

in which μ stands for λH, the extension at the lowest point; therefore

$$ds = c(\sec^2\phi + \mu\sec^3\phi)d\phi,$$
$$\frac{s}{c} = \tan\phi + \frac{1}{2}\mu(\sec\phi\tan\phi + \text{gd}^{-1}\phi), \qquad\text{[prob. 20, p. 37}$$

which is the intrinsic equation of the curve, and reduces to that of the common catenary when $\mu = 0$. The coordinates x, y may be expressed in terms of the single parameter ϕ by putting

$$dx = ds\cos\phi = c(\sec\phi + \mu\sec^2\phi)d\phi,$$
$$dy = ds\sin\phi = c(\sec^2\phi + \mu\sec^3\phi)\sin\phi\, d\phi.$$

ARTICLE 33. THE ELASTIC CATENARY.

Whence

$$\frac{x}{c} = \text{gd}^{-1}\phi + \mu\tan\phi, \quad \frac{y}{c} = \sec\phi + \frac{1}{2}\mu\tan^2\phi.$$

These equations are more convenient than the result of eliminating ϕ, which is somewhat complicated.

Article 34

The Tractory.

[Note.[1]]
To find the equation of the curve which possesses the property that the length of the tangent from the point of contact to the axis of x is constant.

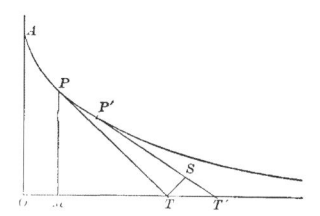

Let PT, $P'T'$ be two consecutive tangents such that $PT = P'T' = c$, and let $OT = t$; draw TS perpendicular to $P'T'$; then if $PP' = ds$, it is evident that ST' differs from ds by an infinitesimal of a higher order. Let PT make an angle ϕ with OA, the axis of y; then (to the first order of infinitesimals) $PT d\phi = TS = TT' \cos\phi$; that is,

$$c\,d\phi = \cos\phi\,dt, \quad t = c\,\mathrm{gd}^{-1}\phi,$$
$$x = t - c\sin\phi = c(\mathrm{gd}^{-1}\phi - \sin\phi), \quad y = c\cos\phi.$$

This is a convenient single-parameter form, which gives all values of x, y as ϕ increases from 0 to $\tfrac{1}{2}\pi$. The value of s, expressed in the same form, is found from the relation

$$ds = ST' = dt\sin\phi = c\tan\phi\,d\phi, \quad s = c\log_e \sec\phi.$$

---

[1] This curve is used in Schiele's anti-friction pivot (Minchin's Statics, Vol. I, p. 242); and in the theory of the skew circular arch, the horizontal projection of the joints being a tractory. (See "Arch," Encyclopædia Britannica.) The equation $\phi = \mathrm{gd}\,\tfrac{t}{c}$ furnishes a convenient method of plotting the curve.

ARTICLE 34. THE TRACTORY.

At the point A, $\phi = 0$, $x = 0$, $s = 0$, $t = 0$, $y = c$. The Cartesian equation, obtained by eliminating ϕ, is

$$\frac{x}{c} = \mathrm{gd}^{-1}\left(\cos^{-1}\frac{y}{c}\right) - \sin\left(\cos^{-1}\frac{y}{c}\right) = \cosh^{-1}\frac{c}{y} - \sqrt{1 - \frac{y^2}{c^2}}.$$

If u be put for $\dfrac{t}{c}$, and be taken as independent variable, $\phi = \mathrm{gd}\,u$, $\dfrac{x}{c} = u - \tanh u$, $\dfrac{y}{c} = \mathrm{sech}\,u$, $\dfrac{s}{c} = \log \cosh u$.

Prob. 100. Given $t = 2c$, show that $\phi = 74° 35'$, $s = 1.3249c$, $y = .2658c$, $x = 1.0360c$. At what point is $t = c$?

Prob. 101. Show that the evolute of the tractory is the catenary. (See Prob. 92.)

Prob. 102. Find the radius of curvature of the tractory in terms of ϕ; and derive the intrinsic equation of the involute.

Article 35

The Loxodrome.

On the surface of a sphere a curve starts from the equator in a given direction and cuts all the meridians at the same angle. To find its equation in latitude-and-longitude coordinates:

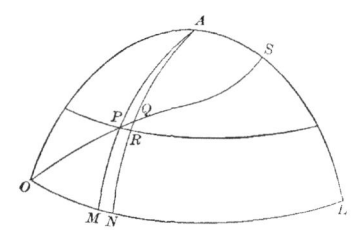

Let the loxodrome cross two consecutive meridians AM, AN in the points P, Q; let PR be a parallel of latitude; let $OM = x$, $MP = y$, $MN' = dx$, $RQ = dy$, all in radian measure; and let the angle $MOP = RPQ = \alpha$; then

$$\tan \alpha = \frac{RQ}{PR}, \text{ but } PR = MN \cos MP,\text{[1]}$$

hence $dx \tan \alpha = dy \sec y$, and $x \tan \alpha = \mathrm{gd}^{-1} y$, there being no integration-constant since y vanishes with x; thus the required equation is

$$y = \mathrm{gd}(x \tan \alpha).$$

To find the length of the arc OP: Integrate the equation

$$ds = dy \csc \alpha, \text{ whence } s = y \csc \alpha.$$

To illustrate numerically, suppose a ship sails northeast, from a point on the equator, until her difference of longitude is 45°, find her latitude and distance:

---
[1] Jones, Trigonometry (Ithaca, 1890), p. 185.

ARTICLE 35. THE LOXODROME.

Here $\tan\alpha = 1$, and $y = \text{gd}\,x = \text{gd}\,\frac{1}{4}\pi = \text{gd}(.7854) = .7152$ radians; $s = y\sqrt{2} = 1.0114$ radii. The latitude in degrees is 40.980.

If the ship set out from latitude y_1, the formula must be modified as follows: Integrating the above differential equation between the limits (x_1, y_1) and (x_2, y_2) gives
$$(x_2 - x_1)\tan\alpha = \text{gd}^{-1} y_2 - \text{gd}^{-1} y_1;$$
hence $\text{gd}^{-1} y_2 = \text{gd}^{-1} y_1 + (x_2 - x_1)\tan\alpha$, from which the final latitude can be found when the initial latitude and the difference of longitude are given. The distance sailed is equal to $(y_2 - y_1)\csc\alpha$ radii, a radius being $60 \times \frac{180}{\pi}$ nautical miles.

Mercator's Chart.—In this projection the meridians are parallel straight lines, and the loxodrome becomes the straight line $y' = x\tan\alpha$, hence the relations between the coordinates of corresponding points on the plane and sphere are $x' = x$, $y' = \text{gd}^{-1} y$. Thus the latitude y is magnified into $\text{gd}^{-1} y$, which is tabulated under the name of "meridional part for latitude y"; the values of y and of y' being given in minutes. A chart constructed accurately from the tables can be used to furnish graphical solutions of problems like the one proposed above.

Prob. 103. Find the distance on a rhumb line between the points (30° N, 20° E) and (30° S, 40° E).

Article 36

Combined Flexure and Tension.

A beam that is built-in at one end carries a load P at the other, and is also subjected to a horizontal tensile force Q applied at the same point; to find the equation of the curve assumed by its neutral surface: Let x, y be any point of the elastic curve, referred to the free end as origin, then the bending moment for this point is $Qy - Px$. Hence, with the usual notation of the theory of flexure,[1]

$$EI\frac{d^2y}{dx^2} = Qy - Px, \quad \frac{d^2y}{dx^2} = n^2(y - mx), \qquad \left[m = \frac{P}{Q}, \; n^2 = \frac{Q}{EI}.\right.$$

which, on putting $y - mx = u$, and $\dfrac{d^2y}{dx^2} = \dfrac{d^2u}{dx^2}$, becomes

$$\frac{d^2u}{dx^2} = n^2 u,$$

whence

$$u = A \cosh nx + B \sinh nx, \qquad \text{[probs. 28, 30}$$

that is,

$$y = mx + A \cosh nx + B \sinh nx.$$

The arbitrary constants A, B are to be determined by the terminal conditions. At the free end $x = 0$, $y = 0$; hence A must be zero, and

$$y = mx + B \sinh nx,$$
$$\frac{dy}{dx} = m + nB \cosh nx;$$

---

[1] Merriman, Mechanics of Materials (New York, 1895), pp. 70–77, 267–269.

ARTICLE 36. COMBINED FLEXURE AND TENSION.

but at the fixed end, $x = l$, and $\dfrac{dy}{dx} = 0$, hence

$$B = -\dfrac{m}{n}\cosh nl,$$

and accordingly

$$y = mx - \dfrac{m \sinh nx}{n \cosh nl}.$$

To obtain the deflection of the loaded end, find the ordinate of the fixed end by putting $x = l$, giving

$$\text{deflection} = m(l - \dfrac{1}{n}\tanh nl),$$

Prob. 104. Compute the deflection of a cast-iron beam, 2×2 inches section, and 6 feet span, built-in at one end and carrying a load of 100 pounds at the other end, the beam being subjected to a horizontal tension of 8000 pounds. [In this case $I = \frac{4}{3}, E = 15 \times 10^6, Q = 8000, P = 100$; hence $n = \frac{1}{50}, m = \frac{1}{80}$, deflection $= \frac{1}{80}(72 - 50\tanh 1.44) = \frac{1}{80}(72 - 44.69) = .341$ inches.]

Prob. 105. If the load be uniformly distributed over the beam, say w per linear unit, prove that the differential equation is

$$EI\dfrac{d^2y}{dx^2} = Qy - \tfrac{1}{2}wx^2, \text{ or } \dfrac{d^2y}{dx^2} = n^2(y - mx^2),$$

and that the solution is $y = A\cosh nx + B\sinh nx + mx^2 + \dfrac{2m}{n^2}$. Show also how to determine the arbitrary constants.

Article 37

Alternating Currents.

[Note.[1]]

In the general problem treated the cable or wire is regarded as having resistance, distributed capacity, self-induction, and leakage; although some of these may be zero in special cases. The line will also be considered to feed into a receiver circuit of any description; and the general solution will include the particular cases in which the receiving end is either grounded or insulated. The electromotive force may, without loss of generality, be taken as a simple harmonic function of the time, because any periodic function can be expressed in a Fourier series of simple harmonics.[2] The E.M.F. and the current, which may differ in phase by any angle, will be supposed to have given values at the terminals of the receiver circuit; and the problem then is to determine the E.M.F. and current that must be kept up at the generator terminals; and also to express the values of these quantities at any intermediate point, distant x from the receiving end; the four line-constants being supposed known, viz.:

$R =$ resistance, in ohms per mile,
$L =$ coefficient of self-induction, in henrys per mile,
$C =$ capacity, in farads per mile,
$G =$ coefficient of leakage, in mhos per mile.[3]

It is shown in standard works[4] that if any simple harmonic function $a\sin(\omega t + \theta)$ be represented by a vector of length a and angle θ, then two simple harmonics of the same period $\frac{2\pi}{\omega}$, but having different values of the phase-angle θ, can be combined by adding their representative vectors. Now the E.M.F. and the current at any point of the circuit, distant x from the receiving end, are of the

---

[1]See references in foot-note Art. 27.
[2]Chapter V, Art. 8.
[3]Kennelly denotes these constants by r, l, c, g. Steinmetz writes s for ωL, κ for ωC, θ for G, and he uses C for current.
[4]Thomson and Tait, Natural Philosophy, Vol. I. p. 40; Rayleigh, Theory of Sound, Vol. I. p. 20; Bedell and Crehore, Alternating Currents, p. 214.

ARTICLE 37. ALTERNATING CURRENTS.

form
$$e = e_1 \sin(\omega t + \theta), \quad i = i_1 \sin(\omega t + \theta'), \tag{64}$$

in which the maximum values e_1, i_1, and the phase-angles θ, θ', are all functions of x. These simple harmonics will be represented by the vectors $e_1\underline{/\theta}$, $i_1\underline{/\theta'}$; whose numerical measures are the complexes $e_1(\cos\theta + j\sin\theta)$[5], $i_1(\cos\theta' + j\sin\theta')$, which will be denoted by $\bar{e}$, $\bar{i}$. The relations between $\bar{e}$ and $\bar{i}$ may be obtained from the ordinary equations[6]

$$\frac{di}{dx} = Ge + C\frac{de}{dt}, \quad \frac{de}{dx} = Ri + L\frac{di}{dt}; \tag{65}$$

for, since $\dfrac{de}{dt} = \omega e_1 \cos(\omega t + \theta) = \omega e_1 \sin(\omega t + \theta + \tfrac{1}{2}\pi)$, then $\dfrac{de}{dt}$ will be represented by the vector $\omega e_1\underline{/\theta + \tfrac{1}{2}\pi}$; and $\dfrac{di}{dx}$ by the sum of the two vectors $Ge_1\underline{/\theta}, C\omega e_1\underline{/\theta + \tfrac{1}{2}\pi}$; whose numerical measures are the complexes $G\bar{e}$, $j\omega C\bar{e}$; and similarly for $\dfrac{de}{dx}$ in the second equation; thus the relations between the complexes $\bar{e}$, $\bar{i}$ are

$$\frac{d\bar{i}}{dx} = (G + j\omega C)\bar{e}, \quad \frac{d\bar{e}}{dx} = (R + j\omega L)\bar{i}. \tag{66}[7]$$

Differentiating and substituting give

$$\left.\begin{array}{l}\dfrac{d^2\bar{e}}{dx^2} = (R + j\omega L)(G + j\omega C)\bar{e},\\[6pt] \dfrac{d^2\bar{i}}{dx^2} = (R + j\omega L)(G + j\omega C)\bar{i},\end{array}\right\} \tag{67}$$

and thus $\bar{e}, \bar{i}$ are similar functions of x, to be distinguished only by their terminal values.

It is now convenient to define two constants m, m_1 by the equations[8]

$$m^2 = (R + j\omega L)(G + j\omega C), \quad m_1 = \frac{m}{(G + j\omega C)}; \tag{68}$$

and the differential equations may then be written

$$\frac{d^2\bar{e}}{dx^2} = m^2\bar{e}, \quad \frac{d^2\bar{i}}{dx^2} = m^2\bar{i}, \tag{69}$$

---

[5]In electrical theory the symbol j is used, instead of i, for $\sqrt{-1}$.

[6]Bedell and Crehore, Alternating Currents, p. 181. The sign of dx is changed, because x is measured from the receiving end. The coefficient of leakage, G, is usually taken zero, but is here retained for generality and symmetry.

[7]These relations have the advantage of not involving the time. Steinmetz derives them from first principles without using the variable t. For instance, he regards $R + j\omega L$ as a generalized resistance-coefficient, which, when applied to i, gives an E.M.F., part of which is in phase with i, and part in quadrature with i. Kennelly calls $R + j\omega L$ the conductor impedance; and $G + j\omega C$ the dielectric admittance; the reciprocal of which is the dielectric impedance.

[8]The complex constants m, m_1 are written z, y by Kennelly; and the variable length x is written L_2. Steinmetz writes v for m.

ARTICLE 37. ALTERNATING CURRENTS.

the solutions of which are[9]

$$\bar{e} = A \cosh mx + B \sinh mx, \quad \bar{\imath} = A' \cosh mx + B' \sinh mx,$$

wherein only two of the four constants are arbitrary; for substituting in either of the equations (66), and equating coefficients, give

$$(G + j\omega C)A = mB', \quad (G + j\omega C)B = mA',$$

whence

$$B' = \frac{A}{m_1}, \quad A' = \frac{B}{m_1}.$$

Next let the assigned terminal values of $\bar{e}, \bar{\imath}$, at the receiver, be denoted by $\bar{E}, \bar{I}$; then putting $x = 0$ gives $\bar{E} = A, \bar{I} = A'$, whence $B = m_1\bar{I}, B' = \dfrac{\bar{E}}{m_1}$; and thus the general solution is

$$\left.\begin{array}{l} \bar{e} = \bar{E} \cosh mx + m_1 \bar{I} \sinh mx, \\[4pt] \bar{\imath} = \bar{I} \cosh mx + \dfrac{\bar{I}}{m_1} \bar{E} \sinh mx. \end{array}\right\} \quad (70)$$

If desired, these expressions could be thrown into the ordinary complex form $X + jY, X' + jY'$, by putting for the letters their complex values, and applying the addition-theorems for the hyperbolic sine and cosine. The quantities X, Y, X', Y' would then be expressed as functions of x; and the representative vectors of e, i, would be $e_1 \underline{/\theta}, i_1 \underline{/\theta'}$, where $e_1^2 = X^2 + Y^2, i^2 = X'^2 + Y'^2, \tan\theta = \dfrac{Y}{X}, \tan\theta' = \dfrac{Y'}{X'}$.

For purposes of numerical computation, however, the formulas (70) are the most convenient, when either a chart,[10] or a table,[11] of $\cosh u, \sinh u$, is available, for complex values of u.

Prob. 106.[12] Given the four line-constants: $R = 2$ ohms per mile, $L = 20$ millihenrys per mile, $C = \frac{1}{2}$ microfarad per mile, $G = 0$; and given ω, the angular velocity of E.M.F. to be 2000 radians per second; then

$$\omega L = 40 \text{ ohms, conductor reactance per mile;}$$
$$R + j\omega L = 2 + 40j \text{ ohms, conductor impedance per mile;}$$
$$\omega C = .001 \text{ mho, dielectric susceptance per mile;}$$
$$G + j\omega C = .001j \text{ mho, dielectric admittance per mile;}$$
$$(G + j\omega C)^{-1} = -1000j \text{ ohms, dielectric impedance per mile;}$$
$$m^2 = (R + j\omega L)(G + j\omega C) = .04 + .002j,$$

which is the measure of $.04005 \underline{/177°8'}$;

---

[9] See Art. 14, Probs. 28–30; and Art. 27, foot-note.
[10] Art. 30, footnote.
[11] See Table II.
[12] The data for this example are taken from Kennelly's article (l. c., p. 38).

ARTICLE 37. ALTERNATING CURRENTS.

therefore

$$m = \text{measure of } .2001\underline{/88°34'} = .0050 + .2000j,$$

an abstract coefficient per mile, of dimensions $[\text{length}]^{-1}$,

$$m_1 = \frac{m}{(G + j\omega C)} = 200 - 5j \text{ ohms}.$$

Next let the assigned terminal conditions at the receiver be: $I = 0$ (line insulated); and $E = 1000$ volts, whose phase may be taken as the standard (or zero) phase; then at any distance x, by (70),

$$\bar{e} = E \cosh mx, \qquad \bar{\imath} = \frac{E}{m_1} \sinh mx,$$

in which mx is an abstract complex.

Suppose it is required to find the E.M.F. and current that must be kept up at a generator 100 miles away; then

$$\bar{e} = 1000 \cosh(.5 + 20j), \quad \bar{\imath} = 200(40 - j)^{-1} \sinh(.5 + 20j),$$

but, by Prob. 89,

$$\cosh(.5 + 20j) = \cosh(.5 + 20j - 6\pi j)$$
$$= \cosh(.5 + 1.15j) = .4600 + .4750j$$

obtained from Table II, by interpolation between $\cosh(.5 + 1.1j)$ and $\cosh(.5 + 1.2j)$; hence

$$\bar{e} = 460 + 475j = e_1(\cos\theta + j\sin\theta),$$

where $\log \tan \theta = \log 475 - \log 460 = .0139$, $\theta = 45° 55'$, and $e_1 = 460 \sec \theta = 661.2$ volts, the required E.M.F.

Similarly $\sinh(.5 + 20j) = \sinh(.5 + 1.15j) = .2126 + 1.0280j$, and hence

$$\bar{\imath} = \frac{200}{1601}(40 + j)(.2126 + 1.028j) = \frac{1}{1601}(1495 + 8266j)$$
$$= i_1(\cos\theta' + j\sin\theta'),$$

where $\log \tan \theta' = 10.7427$, $\theta' = 79° 45'$, $i_1 = 1495 \sec \dfrac{\theta'}{1601} = 5.25$ amperes, the phase and magnitude of required current.

Next let it be required to find e at $x = 8$; then

$$\bar{e} = 1000 \cosh(.04 + 1.6j) = 1000j \sinh(.04 + .03j),$$

by subtracting $\frac{1}{2}\pi j$, and applying page 56. Interpolation between $\sinh(0 + 0j)$ and $\sinh(0 + .1j)$ gives

$$\sinh(0 + .03j) = .00000 + .02995j.$$

Similarly

$$\sinh(.1 + .03j) = .10004 + .03004j.$$

ARTICLE 37. ALTERNATING CURRENTS.

Interpolation between the last two gives

$$\sinh(.04 + .03j) = .04002 + .02999j.$$

Hence $\bar{e} = j(40.02 + 29.99j) = -29.99 + 40.02j = e_1(\cos\theta + j\sin\theta)$, where $\log\tan\theta = .12530$, $\theta = 126° 51'$, $e_1 = -29.99 \sec 126° 51' = 50.01$ volts. Again, let it be required to find e at $x = 16$; here

$$\bar{e} = 1000\cosh(.08 + 3.2j) = -1000\cosh(.08 + .06j),$$

but

$$\cosh(0 + .06j) = .9970 + 0j, \quad \cosh(.1 + .06j) = 1.0020 + .006j;$$

hence

$$\cosh(.08 + .06j) = 1.0010 + .0048j,$$

and

$$\bar{e} = -1001 + 4.8j = e_1(\cos\theta + j\sin\theta),$$

where $\theta = 180° 17'$, $e_1 = 1001$ volts. Thus at a distance of about 16 miles the E.M.F. is the same as at the receiver, but in opposite phase. Since $\bar{e}$ is proportional to $\cosh(.005 + .2j)x$, the value of x for which the phase is exactly $180°$ is $\frac{\pi}{.2} = 15.7$. Similarly the phase of the E.M.F. at $x = 7.85$ is $90°$. There is agreement in phase at any two points whose distance apart is 31.4 miles.

In conclusion take the more general terminal conditions in which the line feeds into a receiver circuit, and suppose the current is to be kept at 50 amperes, in a phase $40°$ in advance of the electromotive force; then $\bar{I}50(\cos 40° + \sin 40°) = 38.30 + 32.14j$, and substituting the constants in (70) gives

$$\bar{c} = 1000\cosh(.005 + .2j)x + (7821 + 6236j)\sinh(.005 + .2j)x$$
$$= 460 + 475j - 4748 + 9366j = -4288 + 9841j = e_1(\cos\theta + j\sin\theta),$$

where $\theta = 113° 33'$, $e_1 = 10730$ volts, the E.M.F. at sending end. This is 17 times what was required when the other end was insulated.

Prob. 107. If $L = 0$, $G = 0$, $I = 0$; then $m = (1+j)n$, $m_1 = (1+j)n_1$ where $n^2 = \dfrac{\omega RC}{2}$, $n_1^2 = \dfrac{R}{2\omega C}$; and the solution is

$$e_1 = \frac{1}{\sqrt{2}} E\sqrt{\cosh 2nx + \cos 2nx}, \qquad \tan\theta = \tan nx \tanh nx,$$

$$i_1 = \frac{1}{2n_1} E\sqrt{\cosh 2nx - \cos 2nx}, \qquad \tan\theta' = \tan nx \coth nx.$$

Prob. 108. If self-induction and capacity be zero, and the receiving end be insulated, show that the graph of the electromotive force is a catenary if $G \neq 0$, a line if $G = 0$.

Prob. 109. Neglecting leakage and capacity, prove that the solution of equations (66) is $\bar{\imath} = \bar{I}$, $\bar{c} = \bar{E} + (R + j\omega L)\bar{I}x$.

ARTICLE 37. ALTERNATING CURRENTS.

Prob. 110. If x be measured from the sending end, show how equations (65), (66) are to be modified; and prove that

$$\bar{e} = \bar{E}_0 \cosh mx - m_1 \bar{I}_0 \sinh mx, \quad \bar{\imath} = \bar{I}_0 \cosh mx - \frac{1}{m_1} \bar{E}_0 \sinh mx,$$

where $\bar{E}_0$, $\bar{I}_0$ refer to the sending end.

Article 38

Miscellaneous Applications.

1. The length of the arc of the logarithmic curve $y = a^x$ is $s = M(\cosh u + \log \tanh \tfrac{1}{2} u)$, in which $M = \dfrac{1}{\log a}$, $\sinh u = \dfrac{y}{M}$.

2. The length of arc of the spiral of Archimedes $r = a\theta$ is $s = \tfrac{1}{4} a(\sinh 2u + 2u)$, where $\sinh u = \theta$.

3. In the hyperbola $\dfrac{x^2}{a^2} - \dfrac{y^2}{b^2} = 1$ the radius of curvature is
$$\rho = \dfrac{(a^2 \sinh^2 u + b^2 \cosh^2 u)^{\frac{3}{2}}}{ab};$$
in which u is the measure of the sector AOP, i.e. $\cosh u = \dfrac{x}{a}$, $\sinh u = \dfrac{y}{b}$.

4. In an oblate spheroid, the superficial area of the zone between the equator and a parallel plane at a distance y is $S = \dfrac{\pi b^2 (\sinh 2u + 2u)}{2e}$, wherein b is the axial radius, e eccentricity, $\sinh u = \dfrac{ey}{p}$, and p parameter of generating ellipse.

5. The length of the arc of the parabola $y^2 = 2px$, measured from the vertex of the curve, is $l = \tfrac{1}{4} p(\sinh 2u + 2u)$, in which $\sinh u = \dfrac{y}{p} = \tan \phi$, where ϕ is the inclination of the terminal tangent to the initial one.

6. The centre of gravity of this arc is given by
$$3l\bar{x} = p^2 (\cosh^3 u - 1), \quad 64 l \bar{y} = p^2 (\sinh 4u - 4u);$$
and the surface of a paraboloid of revolution is $S = 2\pi \bar{y} l$.

7. The moment of inertia of the same arc about its terminal ordinate is $I = \mu \left[xl(x - 2\bar{x}) + \tfrac{1}{64} p^3 N \right]$, where μ is the mass of unit length, and
$$N = u - \dfrac{1}{4} \sinh 2u - \dfrac{1}{4} \sinh 4u + \dfrac{1}{12} \sinh 6u.$$

8. The centre of gravity of the arc of a catenary measured from the lowest point is given by

$$4l\bar{y} = c^2(\sinh 2u + 2u), \quad l\bar{x} = c^2(u\sinh u - \cosh u + 1),$$

in which $u = \frac{x}{c}$; and the moment of inertia of this arc about its terminal abscissa is

$$I = \mu c^3 \left(\frac{1}{12}\sinh 3u + \frac{3}{4}\sinh u - u\cosh u\right).$$

9. Applications to the vibrations of bars are given in Rayleigh, Theory of Sound, Vol. I, art. 170; to the torsion of prisms in Love, Elasticity, pp. 166–74; to the flow of heat and electricity in Byerly, Fourier Series, pp. 75–81; to wave motion in fluids in Rayleigh, Vol. I, Appendix, p. 477, and in Bassett, Hydrodynamics, arts. 120, 384; to the theory of potential in Byerly p. 135, and in Maxwell, Electricity, arts. 172–4; to Non-Euclidian geometry and many other subjects in Günther, Hyperbelfunktionen, Chaps. V and VI. Several numerical examples are worked out in Laisant, Essai sur les fonctions hyperboliques.

Article 39

Explanation of Tables.

In Table I the numerical values of the hyperbolic functions $\sinh u, \cosh u, \tanh u$ are tabulated for values of u increasing from 0 to 4 at intervals of .02. When u exceeds 4, Table IV may be used.

Table II gives hyperbolic functions of complex arguments, in which

$$\cosh(x \pm iy) = a \pm ib, \quad \sinh(x \pm iy) = c \pm id,$$

and the values of a, b, c, d are tabulated for values of x and of y ranging separately from 0 to 1.5 at intervals of .1. When interpolation is necessary it may be performed in three stages. For example, to find $\cosh(.82 + 1.34i)$: First find $\cosh(.82+1.3i)$, by keeping y at 1.3 and interpolating between the entries under $x = .8$ and $x = .9$; next find $\cosh(.82+1.4i)$, by keeping y at 1.4 and interpolating between the entries under $x = .8$ and $x = .9$, as before; then by interpolation between $\cosh(.82 + 1.3i)$ and $\cosh(.82 + 1.4i)$ find $\cosh(.82 + 1.34i)$, in which x is kept at .82. The table is available for all values of y, however great, by means of the formulas on page 56:

$$\sinh(x + 2i\pi) = \sinh x, \quad \cosh(x + 2i\pi) = \cosh x, \text{ etc.}$$

It does not apply when x is greater than 1.5, but this case seldom occurs in practice. This table can also be used as a complex table of circular functions, for

$$\cos(y \pm ix) = a \mp ib, \quad \sin(y \pm ix) = d \pm ic;$$

and, moreover, the exponential function is given by

$$\exp(\pm x \pm iy) = a \pm c \pm i(b \pm d),$$

in which the signs of c and d are to be taken the same as the sign of x, and the sign of i on the right is to be the product of the signs of x and of i on the left.

Table III gives the values of $v = \operatorname{gd} u$, and of the gudermanian angle $\theta = \dfrac{180°v}{\pi}$, as u changes from 0 to 1 at intervals of .02, from 1 to 2 at intervals of .05, and from 2 to 4 at intervals of .1.

ARTICLE 39. EXPLANATION OF TABLES. 80

In Table IV are given, the values of gd u, log sinh u, log cosh u, as u increases from 4 to 6 at intervals of .1, from 6 to 7 at intervals of .2, and from 7 to 9 at intervals of .5.

In the rare cases in which more extensive tables are necessary, reference may be made to the tables[1] of Gudermann, Glaisher, and Geipel and Kilgour. In the first the Gudermanian angle (written k) is taken as the independent variable, and increases from 0 to 100 grades at intervals of .01, the corresponding value of u (written Lk) being tabulated. In the usual case, in which the table is entered with the value of u, it gives by interpolation the value of the gudermanian angle, whose circular functions would then give the hyperbolic functions of u. When u is large, this angle is so nearly right that interpolation is not reliable. To remedy this inconvenience Gudermann's second table gives directly log sinh u, log cosh u, log tanh u, to nine figures, for values of u varying by .001 from 2 to 5, and by .01 from 5 to 12.

Glaisher has tabulated the values of e^x and e^{-x}, to nine significant figures, as x varies by .001 from 0 to .1, by .01 from 0 to 2, by .1 from 0 to 10, and by 1 from 0 to 500. From these the values of cosh x, sinh x are easily obtained.

Geipel and Kilgour's handbook gives the values of cosh x, sinh x, to seven figures, as x varies by .01 from 0 to 4.

There are also extensive tables by Forti, Gronau, Vassal, Callet, and Hoüel; and there are four-place tables in Byerly's Fourier Series, and in Wheeler's Trigonometry.

In the following tables a dash over a final digit indicates that the number has been increased.

---

[1] Gudermann in Crelle's Journal, vols. 6–9, 1831–2 (published separately under the title Theorie der hyperbolischen Functionen, Berlin, 1833). Glaisher in Cambridge Phil. Trans., vol. 13, 1881. Geipel and Kilgour's Electrical Handbook.

Table I.—Hyperbolic Functions.

u	$\sinh u$	$\cosh u$	$\tanh u$	u	$\sinh u$	$\cosh u$	$\tanh u$
.00	.0000	1.0000	.0000	1.00	1.1752	1.543$\bar{1}$	.7616
02	0200	1.0002	0200	1.02	1.206$\bar{3}$	1.566$\bar{9}$	769$\bar{9}$
04	0400	1.0008	040$\bar{0}$	1.04	1.237$\bar{9}$	1.5913	777$\bar{9}$
06	0600	1.0018	0599	1.06	1.270$\bar{0}$	1.6164	785$\bar{7}$
08	080$\bar{1}$	1.0032	0798	1.08	1.3025	1.6421	793$\bar{2}$
.10	.100$\bar{2}$	1.0050	.099$\bar{7}$	1.10	1.3356	1.6685	.8005
12	120$\bar{3}$	1.0072	1194	1.12	1.369$\bar{3}$	1.695$\bar{6}$	807$\bar{6}$
14	140$\bar{5}$	1.0098	139$\bar{1}$	1.14	1.403$\bar{5}$	1.7233	8144
16	160$\bar{7}$	1.0128	1586	1.16	1.4382	1.7517	8210
18	181$\bar{0}$	1.0162	178$\bar{1}$	1.18	1.4735	1.7808	827$\bar{5}$
.20	.2013	1.020$\bar{1}$	.197$\bar{4}$	1.20	1.509$\bar{5}$	1.810$\bar{7}$	.833$\bar{7}$
22	221$\bar{8}$	1.024$\bar{3}$	2165	1.22	1.546$\bar{0}$	1.8412	839$\bar{7}$
24	2423	1.0289	235$\bar{5}$	1.24	1.5831	1.872$\bar{5}$	845$\bar{5}$
26	2629	1.034$\bar{0}$	254$\bar{3}$	1.26	1.620$\bar{9}$	1.9045	851$\bar{1}$
28	283$\bar{7}$	1.0395	2729	1.28	1.6593	1.9373	856$\bar{5}$
.30	.3045	1.0453	.2913	1.30	1.6984	1.9709	.8617
32	325$\bar{5}$	1.0516	3095	1.32	1.7381	2.005$\bar{3}$	8668
34	3466	1.058$\bar{4}$	327$\bar{5}$	1.34	1.778$\bar{6}$	2.0404	871$\bar{7}$
36	3678	1.0655	3452	1.36	1.819$\bar{8}$	2.0764	876$\bar{4}$
38	3892	1.0731	3627	1.38	1.861$\bar{7}$	2.1132	881$\bar{0}$
.40	.410$\bar{8}$	1.081$\bar{1}$	.3799	1.40	1.9043	2.150$\bar{9}$	.8854
42	432$\bar{5}$	1.0895	3969	1.42	1.9477	2.1894	889$\bar{6}$
44	4543	1.098$\bar{4}$	4136	1.44	1.991$\bar{9}$	2.2288	893$\bar{7}$
46	476$\bar{4}$	1.107$\bar{7}$	430$\bar{1}$	1.46	2.036$\bar{9}$	2.269$\bar{1}$	897$\bar{7}$
48	4986	1.1174	4462	1.48	2.082$\bar{7}$	2.310$\bar{3}$	901$\bar{5}$
.50	.521$\bar{1}$	1.1276	.4621	1.50	2.129$\bar{3}$	2.3524	.9051
52	543$\bar{8}$	1.138$\bar{3}$	4777	1.52	2.176$\bar{8}$	2.395$\bar{5}$	908$\bar{7}$
54	5666	1.149$\bar{4}$	493$\bar{0}$	1.54	2.2251	2.439$\bar{5}$	9121
56	5897	1.1609	508$\bar{0}$	1.56	2.2743	2.484$\bar{5}$	9154
58	613$\bar{1}$	1.173$\bar{0}$	522$\bar{7}$	1.58	2.324$\bar{5}$	2.530$\bar{5}$	9186
.60	.636$\bar{7}$	1.185$\bar{5}$	.5370	1.60	2.375$\bar{6}$	2.577$\bar{5}$	.921$\bar{7}$
62	660$\bar{5}$	1.1984	5511	1.62	2.427$\bar{6}$	2.625$\bar{5}$	9246
64	684$\bar{6}$	1.211$\bar{9}$	564$\bar{9}$	1.64	2.480$\bar{6}$	2.674$\bar{6}$	927$\bar{5}$
66	709$\bar{0}$	1.2258	578$\bar{4}$	1.66	2.534$\bar{6}$	2.7247	9302
68	7336	1.2402	5915	1.68	2.5896	2.776$\bar{0}$	932$\bar{9}$
.70	.758$\bar{6}$	1.255$\bar{2}$	.604$\bar{4}$	1.70	2.6456	2.8283	.9354
72	7838	1.270$\bar{6}$	6169	1.72	2.7027	2.881$\bar{8}$	937$\bar{9}$
74	8094	1.2865	6291	1.74	2.7609	2.9364	9402
76	8353	1.303$\bar{0}$	641$\bar{1}$	1.76	2.820$\bar{2}$	2.9922	9425
78	8615	1.3199	6527	1.78	2.8806	3.0492	944$\bar{7}$
.80	.8881	1.3374	.6640	1.80	2.942$\bar{2}$	3.107$\bar{5}$	.9468
82	9150	1.355$\bar{5}$	675$\bar{1}$	1.82	3.0049	3.1669	9488
84	9423	1.3740	6858	1.84	3.068$\bar{9}$	3.227$\bar{7}$	950$\bar{8}$
86	970$\bar{0}$	1.393$\bar{2}$	696$\bar{3}$	1.86	3.1340	3.2897	952$\bar{7}$
88	998$\bar{1}$	1.4128	7064	1.88	3.200$\bar{5}$	3.3530	954$\bar{5}$
.90	1.0265	1.433$\bar{1}$	.716$\bar{3}$	1.90	3.268$\bar{2}$	3.4177	.9562
92	1.0554	1.4539	725$\bar{9}$	1.92	3.337$\bar{2}$	3.483$\bar{8}$	9579
94	1.084$\bar{7}$	1.4753	7352	1.94	3.4075	3.5512	9595
96	1.1144	1.497$\bar{3}$	744$\bar{3}$	1.96	3.4792	3.620$\bar{1}$	961$\bar{1}$
98	1.144$\bar{6}$	1.519$\bar{9}$	753$\bar{1}$	1.98	3.5523	3.6904	962$\bar{6}$

TABLE I.—HYPERBOLIC FUNCTIONS (*continued*)

u	sinh u.	cosh u.	tanh u.	u	sinh u.	cosh u.	tanh u.
2.00	3.626$\bar{9}$	3.762$\bar{2}$	.9640	3.00	10.017$\bar{9}$	10.067$\bar{7}$	.99505
2.02	3.7028	3.835$\bar{5}$	9654	3.02	10.2212	10.2700	99524
2.04	3.780$\bar{3}$	3.9103	9667	3.04	10.4287	10.4765	99543
2.06	3.859$\bar{3}$	3.9867	9680	3.06	10.6403	10.6872	99561
2.08	3.9398	4.0647	969$\bar{3}$	3.08	10.8562	10.902$\bar{2}$	99578
2.10	4.021$\bar{9}$	4.1443	.970$\bar{5}$	3.10	11.076$\bar{5}$	11.1215	.99594
2.12	4.1055	4.225$\bar{6}$	971$\bar{6}$	3.12	11.3011	11.345$\bar{3}$	99610
2.14	4.190$\bar{9}$	4.3085	972$\bar{7}$	3.14	11.530$\bar{3}$	11.573$\bar{6}$	99626
2.16	4.2779	4.3932	9737	3.16	11.764$\bar{1}$	11.8065	99640
2.18	4.3666	4.479$\bar{7}$	974$\bar{8}$	3.18	12.002$\bar{6}$	12.044$\bar{2}$	99654
2.20	4.4571	4.5679	.9757	3.20	12.245$\bar{9}$	12.2866	.99668
2.22	4.549$\bar{4}$	4.658$\bar{0}$	976$\bar{7}$	3.22	12.494$\bar{1}$	12.5340	99681
2.24	4.6434	4.749$\bar{9}$	977$\bar{6}$	3.24	12.747$\bar{3}$	12.7864	99693
2.26	4.739$\bar{4}$	4.8437	978$\bar{5}$	3.26	13.005$\bar{6}$	13.044$\bar{0}$	99705
2.28	4.837$\bar{2}$	4.939$\bar{5}$	979$\bar{3}$	3.28	13.269$\bar{1}$	13.3067	99717
2.30	4.937$\bar{0}$	5.0372	.980$\bar{1}$	3.30	13.537$\bar{9}$	13.574$\bar{8}$	.99728
2.32	5.0387	5.137$\bar{0}$	980$\bar{9}$	3.32	13.812$\bar{1}$	13.848$\bar{3}$	99738
2.34	5.1425	5.238$\bar{8}$	9816	3.34	14.0918	14.127$\bar{3}$	99749
2.36	5.248$\bar{3}$	5.342$\bar{7}$	9823	3.36	14.3772	14.412$\bar{0}$	99758
2.38	5.356$\bar{2}$	5.4487	9830	3.38	14.668$\bar{4}$	14.7024	99768
2.40	5.4662	5.5569	.983$\bar{7}$	3.40	14.965$\bar{4}$	14.9987	.99777
2.42	5.5785	5.667$\bar{4}$	9843	3.42	15.268$\bar{4}$	15.301$\bar{1}$	99786
2.44	5.6929	5.7801	9849	3.44	15.5774	15.6095	99794
2.46	5.809$\bar{7}$	5.8951	9855	3.46	15.892$\bar{8}$	15.9242	99802
2.48	5.928$\bar{8}$	6.0125	986$\bar{1}$	3.48	16.2144	16.245$\bar{3}$	99810
2.50	6.0502	6.132$\bar{3}$	.9866	3.50	16.5426	16.5728	.99817
2.52	6.174$\bar{1}$	6.2545	9871	3.52	16.8774	16.9070	99824
2.54	6.3004	6.379$\bar{3}$	9876	3.54	17.219$\bar{0}$	17.248$\bar{0}$	99831
2.56	6.429$\bar{3}$	6.506$\bar{6}$	9881	3.56	17.567$\bar{4}$	17.5958	99831
2.58	6.560$\bar{7}$	6.6364	988$\bar{6}$	3.58	17.9228	17.9507	99844
2.60	6.6947	6.7690	.9890	3.60	18.2854	18.312$\bar{8}$	.99850
2.62	6.831$\bar{5}$	6.904$\bar{3}$	989$\bar{5}$	3.62	18.655$\bar{4}$	18.682$\bar{2}$	99856
2.64	6.9709	7.042$\bar{3}$	989$\bar{9}$	3.64	19.032$\bar{8}$	19.0590	99862
2.66	7.113$\bar{2}$	7.183$\bar{2}$	990$\bar{3}$	3.66	19.4178	19.4435	99867
2.68	7.258$\bar{3}$	7.3268	9906	3.68	19.810$\bar{6}$	19.8358	99872
2.70	7.406$\bar{3}$	7.473$\bar{5}$	.9910	3.70	20.211$\bar{3}$	20.2360	.99877
2.72	7.5572	7.623$\bar{1}$	991$\bar{4}$	3.72	20.620$\bar{1}$	20.6443	99882
2.74	7.7112	7.775$\bar{8}$	991$\bar{7}$	3.74	21.0371	21.060$\bar{9}$	99887
2.76	7.868$\bar{3}$	7.931$\bar{6}$	9920	3.76	21.462$\bar{6}$	21.485$\bar{9}$	99891
2.78	8.028$\bar{5}$	8.0905	9923	3.78	21.8966	21.9194	9989$\bar{6}$
2.80	8.1919	8.2527	.9926	3.80	22.3394	22.361$\bar{8}$	.9990$\bar{0}$
2.82	8.3586	8.4182	9929	3.82	22.7911	22.813$\bar{1}$	9990$\bar{4}$
2.84	8.528$\bar{7}$	8.587$\bar{1}$	993$\bar{2}$	3.84	23.252$\bar{0}$	23.273$\bar{5}$	99907
2.86	8.7021	8.759$\bar{4}$	993$\bar{5}$	3.86	23.7721	23.7432	99911
2.88	8.879$\bar{1}$	8.9352	9937	3.88	24.2018	24.2224	9991$\bar{5}$
2.90	9.059$\bar{6}$	9.114$\bar{6}$	.994$\bar{0}$	3.90	24.6911	24.7113	.99918
2.92	9.243$\bar{7}$	9.2976	994$\bar{2}$	3.92	25.1903	25.2101	99921
2.94	9.431$\bar{5}$	9.484$\bar{4}$	9944	3.94	25.699$\bar{6}$	25.7190	99924
2.96	9.623$\bar{1}$	9.674$\bar{9}$	994$\bar{7}$	3.96	26.2191	26.238$\bar{2}$	99927
2.98	9.8185	9.8693	994$\bar{9}$	3.98	26.749$\bar{2}$	26.767$\bar{9}$	99930

TABLE II.—VALUES OF $\cosh(x+iy)$ AND $\sinh(x+iy)$.

	$x = 0$				$x = .1$			
y	a	b	c	d	a	b	c	d
0	1.0000	0000	0000	.0000	1.0050	.00000	.100$\bar{1}\bar{7}$	.0000
.1	0.9950	"	"	0998	1.000$\bar{0}$	01000	09967	1003
.2	0.980$\bar{1}$	"	"	198$\bar{7}$	0.9850	0199$\bar{0}$	09817	199$\bar{7}$
.3	0.9553	"	"	2955	0.9601	02960	0957$\bar{0}$	297$\bar{0}$
.4	.921$\bar{1}$	"	"	.3894	.925$\bar{7}$	.03901	.09226	.3914
.5	8776	"	"	4794	882$\bar{0}$	04802	0879$\bar{1}$	4818
.6	8253	"	"	5646	829$\bar{5}$	05656	08267	567$\bar{5}$
.7	7648	"	"	6442	768$\bar{7}$	06453	07661	6474
.8	.6967	"	"	.717$\bar{4}$	.700$\bar{2}$	.0718$\bar{6}$	.0697$\bar{9}$	.7800
.9	6216	"	"	7833	624$\bar{7}$	078$\bar{4}\bar{7}$	0622$\bar{7}$	7872
1.0	5403	"	"	841$\bar{5}$	5430	08429	05412	845$\bar{7}$
1.1	4536	"	"	8912	455$\bar{9}$	08927	04544	895$\bar{7}$
1.2	.362$\bar{4}$	"	"	.9320	.364$\bar{2}$	.09336	.0363$\bar{0}$	0.936$\bar{7}$
1.3	2675	"	"	963$\bar{6}$	268$\bar{8}$	096$\bar{5}\bar{2}$	0268$\bar{0}$	0.968$\bar{4}$
1.4	170$\bar{0}$	"	"	9854	1708	09871	0170$\bar{3}$	0.990$\bar{4}$
1.5	0707	"	"	997$\bar{5}$	0711	0999$\bar{2}$	0070$\bar{9}$	1.002$\bar{5}$
$\frac{1}{2}\pi$	0000	"	"	1.0000	0000	100$\bar{1}\bar{7}$	00000	1.0050

	$x = .2$				$x = .3$			
y	a	b	c	d	a	b	c	d
0	1.020$\bar{1}$	.0000	.2013	.0000	1.0453	.0000	.3045	.0000
.1	1.015$\bar{0}$	0201	2003	1018	1.040$\bar{1}$	0304	303$\bar{0}$	1044
.2	0.9997	0400	1973	202$\bar{7}$	1.024$\bar{5}$	0605	298$\bar{5}$	207$\bar{7}$
.3	0.9745	0595	1923	3014	9987	090$\bar{0}$	2909	3089
.4	.9395	.0784	.1854	.3972	.9628	.1186	.280$\bar{5}$	.407$\bar{1}$
.5	895$\bar{2}$	0965	176$\bar{7}$	4890	917$\bar{4}$	146$\bar{0}$	267$\bar{2}$	501$\bar{2}$
.6	8419	113$\bar{7}$	166$\bar{2}$	576$\bar{0}$	8687	1719	2513	590$\bar{3}$
.7	780$\bar{2}$	1297	154$\bar{0}$	6571	7995	196$\bar{2}$	2329	6734
.8	.710$\bar{7}$	.1444	.140$\bar{3}$	.731$\bar{8}$	.728$\bar{3}$	.2184	.212$\bar{2}$	.7498
.9	634$\bar{1}$	1577	125$\bar{2}$	7990	649$\bar{8}$	2385	189$\bar{3}$	8188
1.0	5511	1694	108$\bar{8}$	858$\bar{4}$	5648	2562	1645	8796
1.1	4627	179$\bar{5}$	0913	909$\bar{1}$	474$\bar{2}$	2714	1381	9316
1.2	.3696	.187$\bar{7}$	.073$\bar{0}$	0.9507	.378$\bar{8}$	.2838	.1103	0.974$\bar{3}$
1.3	272$\bar{9}$	1940	053$\bar{9}$	0.982$\bar{9}$	2796	2934	081$\bar{5}$	1.0072
1.4	173$\bar{4}$	1984	0342	1.0052	177$\bar{7}$	3001	051$\bar{8}$	1.0301
1.5	072$\bar{2}$	2008	0142	1.0175	0739	303$\bar{8}$	0215	1.042$\bar{7}$
$\frac{1}{2}\pi$	0000	2013	0000	1.020$\bar{1}$	0000	3045	0000	1.0453

TABLE II.—VALUES OF $\cosh(x+iy)$ AND $\sinh(x+iy)$. (continued)

	\multicolumn{4}{c	}{$x=.4$}	\multicolumn{4}{c}{$x=.5$}
y	a	b	c	d	a	b	c	d
0	1.081$\bar{1}$	.0000	.410$\bar{8}$	.0000	1.1276	.0000	.521$\bar{1}$	.0000
.1	1.0756	0410	408$\bar{7}$	1079	1.122$\bar{0}$	0520	518$\bar{5}$	1126
.2	1.0595	0816	402$\bar{6}$	214$\bar{8}$	1.1051	1025	5107	2240
.3	1.032$\bar{8}$	121$\bar{4}$	3924	319$\bar{5}$	1.077$\bar{3}$	154$\bar{0}$	4978	3332
.4	.9957	.160$\bar{0}$	.3783	.421$\bar{0}$	1.0386	.2029	.480$\bar{0}$	.4391
.5	9487	1969	360$\bar{5}$	518$\bar{3}$	0.989$\bar{6}$	2498	4573	5406
.6	8922	2319	3390	6104	0.9306	2942	430$\bar{1}$	6367
.7	8268	2646	314$\bar{2}$	6964	0.8624	335$\bar{7}$	398$\bar{6}$	7264
.8	.753$\bar{2}$	.2947	.286$\bar{2}$	.7755	.7856	.3738	.363$\bar{1}$	0.8089
.9	672$\bar{0}$	3218	2553	8468	7009	408$\bar{2}$	3239	0.8833
1.0	5841	3456	2219	909$\bar{7}$	609$\bar{3}$	438$\bar{5}$	2815	0.948$\bar{9}$
1.1	4904	366$\bar{1}$	1863	963$\bar{5}$	511$\bar{5}$	4644	236$\bar{4}$	1.005$\bar{0}$
1.2	.3917	.328$\bar{9}$	.1488	1.0076	.4056	.485$\bar{7}$	.1888	1.051$\bar{0}$
1.3	289$\bar{2}$	395$\bar{8}$	109$\bar{9}$	1.041$\bar{7}$	3016	5021	139$\bar{4}$	1.0865
1.4	183$\bar{8}$	404$\bar{8}$	0698	1.0653	191$\bar{7}$	513$\bar{5}$	088$\bar{6}$	1.1163
1.5	076$\bar{5}$	4097	029$\bar{1}$	1.078$\bar{4}$	079$\bar{8}$	519$\bar{8}$	036$\bar{9}$	1.124$\bar{8}$
$\frac{1}{2}\pi$	0000	410$\bar{8}$	0000	1.081$\bar{1}$	0000	521$\bar{1}$	0000	1.1276

	\multicolumn{4}{c	}{$x=.6$}	\multicolumn{4}{c}{$x=.7$}
y	a	b	c	d	a	b	c	d
0	1.185$\bar{5}$	.0000	.636$\bar{7}$	.0000	1.2552	.0000	.758$\bar{6}$	.0000
.1	1.1795	063$\bar{6}$	633$\bar{5}$	1183	1.248$\bar{9}$	0757	754$\bar{8}$	1253
.2	1.161$\bar{8}$	126$\bar{5}$	624$\bar{0}$	2355	1.2301	1542	743$\bar{5}$	249$\bar{4}$
.3	1.132$\bar{5}$	1881	6082	3503	1.1991	224$\bar{2}$	7247	3709
.4	1.0918	.2479	.5864	.461$\bar{7}$	1.156$\bar{1}$	.2954	.6987	.488$\bar{8}$
.5	1.0403	3052	5587	5684	1.1015	363$\bar{7}$	6657	601$\bar{8}$
.6	0.9784	395$\bar{5}$	525$\bar{5}$	669$\bar{4}$	1.0359	4253	626$\bar{1}$	7087
.7	0.906$\bar{7}$	4101	4869	763$\bar{7}$	0.960$\bar{0}$	488$\bar{7}$	580$\bar{2}$	8086
.8	.8259	.4567	.443$\bar{6}$	0.8504	.874$\bar{5}$	.544$\bar{2}$	.5285	0.9004
.9	736$\bar{9}$	4987	3957	0.9286	7802	5942	4715	0.9832
1.0	6405	5357	344$\bar{0}$	0.9975	678$\bar{2}$	6383	409$\bar{9}$	1.056$\bar{2}$
1.1	5377	567$\bar{4}$	2888	1.056$\bar{5}$	5693	6760	344$\bar{1}$	1.1186
1.2	.429$\bar{6}$	.593$\bar{4}$	.230$\bar{7}$	1.104$\bar{9}$	.4548	.7070	.274$\bar{9}$	1.169$\bar{9}$
1.3	3171	613$\bar{5}$	1703	1.1422	335$\bar{8}$	7309	2029	1.2094
1.4	201$\bar{5}$	627$\bar{4}$	1082	1.1682	2133	7475	1289	1.2369
1.5	083$\bar{9}$	635$\bar{1}$	0450	1.182$\bar{5}$	088$\bar{8}$	756$\bar{7}$	0537	1.2520
$\frac{1}{2}\pi$	0000	636$\bar{7}$	0000	1.185$\bar{5}$	0000	758$\bar{6}$	0000	1.2552

TABLE II.—Values of $\cosh(x+iy)$ and $\sinh(x+iy)$. (continued)

	\multicolumn{4}{c	}{$x = .8$}	\multicolumn{4}{c}{$x = .9$}
y	a	b	c	d	a	b	c	d
0	1.3374	.0000	.8881	.0000	1.433$\bar{1}$	.0000	1.0265	.0000
.1	1.330$\bar{8}$	088$\bar{7}$	883$\bar{7}$	1335	1.4259	102$\bar{5}$	1.021$\bar{4}$	143$\bar{1}$
.2	1.3108	1764	8704	2657	1.4045	2039	1.006$\bar{1}$	2847
.3	1.2776	262$\bar{5}$	8484	3952	1.3691	303$\bar{4}$	0.980$\bar{7}$	4235
.4	1.231$\bar{9}$	.3458	.8180	.5208	1.320$\bar{0}$	.3997	.945$\bar{5}$	.558$\bar{1}$
.5	1.173$\bar{7}$	425$\bar{8}$	779$\bar{4}$	641$\bar{2}$	1.257$\bar{7}$	4921	9008	687$\bar{1}$
.6	1.1038	501$\bar{5}$	733$\bar{0}$	755$\bar{2}$	1.182$\bar{8}$	5796	8472	809$\bar{2}$
.7	1.0229	5721	679$\bar{3}$	861$\bar{6}$	1.096$\bar{1}$	661$\bar{3}$	7851	9232
.8	.931$\bar{8}$	.637$\bar{1}$	.618$\bar{8}$	0.9595	.9984	.736$\bar{4}$	.715$\bar{2}$	1.0280
.9	831$\bar{4}$	695$\bar{7}$	552$\bar{1}$	1.0476	8908	804$\bar{1}$	638$\bar{1}$	1.1226
1.0	7226	7472	4798	1.1254	7743	8638	5546	1.205$\bar{9}$
1.1	606$\bar{7}$	791$\bar{5}$	4028	1.1919	6500	9148	4656	1.277$\bar{2}$
1.2	.4846	.827$\bar{8}$	.3218	1.2465	.519$\bar{3}$	0.956$\bar{8}$	.372$\bar{0}$	1.335$\bar{7}$
1.3	357$\bar{8}$	8557	237$\bar{6}$	1.288$\bar{7}$	383$\bar{4}$	0.9891	274$\bar{6}$	1.380$\bar{9}$
1.4	2273	875$\bar{2}$	151$\bar{0}$	1.3180	2436	1.0124	1745	1.4122
1.5	0946	885$\bar{9}$	0628	1.334$\bar{1}$	101$\bar{4}$	1.0239	0726	1.429$\bar{5}$
$\tfrac{1}{2}\pi$	0000	.8881	0000	1.3374	0000	1.0265	0000	1.433$\bar{1}$

	\multicolumn{4}{c	}{$x = 1.0$}	\multicolumn{4}{c}{$x = 1.1$}
y	a	b	c	d	a	b	c	d
0	1.543$\bar{1}$	.0000	1.1752	.0000	1.6685	.0000	1.3356	.0000
.1	1.535$\bar{4}$	1173	1.1693	154$\bar{1}$	1.660$\bar{2}$	1333	1.329$\bar{0}$	1666
.2	1.5123	2335	1.1518	306$\bar{6}$	1.635$\bar{3}$	2654	1.3090	331$\bar{5}$
.3	1.474$\bar{2}$	347$\bar{3}$	1.1227	4560	1.594$\bar{0}$	3946	1.276$\bar{0}$	493$\bar{1}$
.4	1.421$\bar{3}$	457$\bar{6}$	1.0824	.6009	1.5368	5201	1.2302	0.649$\bar{8}$
.5	1.354$\bar{2}$	5634	1.031$\bar{4}$	739$\bar{8}$	1.464$\bar{3}$	6403	1.1721	0.7999
.6	1.273$\bar{6}$	663$\bar{6}$	0.9699	871$\bar{8}$	1.377$\bar{1}$	754$\bar{2}$	1.102$\bar{4}$	0.9421
.7	1.1802	757$\bar{1}$	0.8988	994$\bar{1}$	1.276$\bar{2}$	8604	1.021$\bar{6}$	1.074$\bar{9}$
.8	1.075$\bar{1}$	0.8430	.818$\bar{8}$	1.1069	1.162$\bar{5}$	0.9581	.930$\bar{6}$	1.1969
.9	0.9592	0.920$\bar{6}$	7305	1.2087	1.037$\bar{2}$	1.0462	8302	1.3070
1.0	0.8337	0.9889	635$\bar{0}$	1.298$\bar{5}$	0.9015	1.1239	721$\bar{7}$	1.4040
1.1	0.6999	1.0473	533$\bar{1}$	1.375$\bar{2}$	0.7568	1.1903	6058	1.487$\bar{0}$
1.2	.559$\bar{2}$	1.0953	.4258	1.4382	.6046	1.244$\bar{9}$	.484$\bar{0}$	1.5551
1.3	5128	1.132$\bar{4}$	314$\bar{4}$	1.486$\bar{8}$	4463	1.287$\bar{0}$	357$\bar{5}$	1.6077
1.4	262$\bar{3}$	1.158$\bar{1}$	199$\bar{8}$	1.5213	2836	1.3162	2270	1.6442
1.5	109$\bar{2}$	1.172$\bar{3}$	0831	1.5392	1180	1.332$\bar{3}$	094$\bar{5}$	1.6643
$\tfrac{1}{2}\pi$	0000	1.1752	0000	1.543$\bar{1}$	.0000	1.3356	.0000	1.6685

Table II.—Values of $\cosh(x+iy)$ and $\sinh(x+iy)$. (continued)

	\multicolumn{4}{c}{$x = 1.2$}	\multicolumn{4}{c}{$x = 1.3$}
y	a	b	c	d	a	b	c	d
0	1.810$\overline{7}$	.0000	1.509$\overline{5}$	.0000	1.9709	0000	1.698$\overline{4}$	.0000
.1	1.8016	150$\overline{7}$	1.5019	180$\overline{8}$	1.961$\overline{1}$	169$\overline{6}$	1.689$\overline{9}$	196$\overline{8}$
.2	1.774$\overline{6}$	299$\overline{9}$	1.479$\overline{4}$	359$\overline{8}$	1.9316	3374	1.6645	3916
.3	1.729$\overline{8}$	446$\overline{1}$	1.4420	535$\overline{1}$	1.882$\overline{9}$	5019	1.6225	5824
.4	1.6677	.5878	1.3903	0.7051	1.8153	.661$\overline{4}$	1.5643	0.7675
.5	1.5890	723$\overline{7}$	1.324$\overline{7}$	0.868$\overline{1}$	1.7296	8142	1.490$\overline{5}$	0.9449
.6	1.4944	8523	1.2458	1.022$\overline{4}$	1.626$\overline{7}$	959$\overline{0}$	1.4017	1.1131
.7	1.384$\overline{9}$	9724	1.154$\overline{5}$	1.166$\overline{5}$	1.5074	1.0941	1.299$\overline{0}$	1.2697
.8	1.261$\overline{5}$	1.0828	1.051$\overline{7}$	1.298$\overline{9}$	1.3731	1.2183	1.183$\overline{3}$	1.413$\overline{9}$
.9	1.1255	1.182$\overline{4}$	0.938$\overline{3}$	1.4183	1.2251	1.330$\overline{4}$	1.0557	1.543$\overline{9}$
1.0	0.9783	1.270$\overline{2}$	0.815$\overline{6}$	1.5236	1.064$\overline{9}$	1.4291	0.9176	1.658$\overline{5}$
1.1	0.8213	1.3452	0.684$\overline{7}$	1.613$\overline{7}$	0.8940	1.5136	0.770$\overline{4}$	1.756$\overline{5}$
1.2	.6561	1.406$\overline{9}$	0.547$\overline{0}$	1.6876	.714$\overline{2}$	1.583$\overline{0}$	0.6154	1.837$\overline{0}$
1.3	484$\overline{4}$	1.4544	0.403$\overline{8}$	1.7447	5272	1.636$\overline{5}$	0.4543	1.899$\overline{1}$
1.4	307$\overline{8}$	1.4875	0.256$\overline{6}$	1.7843	3350	1.673$\overline{7}$	0.288$\overline{7}$	1.9422
1.5	128$\overline{1}$	1.505$\overline{7}$	0.106$\overline{8}$	1.8061	1394	1.6941	0.1201	1.966$\overline{0}$
$\tfrac{1}{2}\pi$	0000	1.509$\overline{5}$	0000	1.810$\overline{7}$	0000	1.698$\overline{4}$	0000	1.9709

	\multicolumn{4}{c}{$x = 1.4$}	\multicolumn{4}{c}{$x = 1.5$}
y	a	b	c	d	a	b	c	d
0	2.150$\overline{9}$	.0000	1.9043	.0000	2.3524	.0000	2.129$\overline{3}$	.0000
.1	2.1401	1901	1.8948	2147	2.3413	2126	2.118$\overline{7}$	2348
.2	2.1080	3783	1.8663	4273	2.3055	4230	2.0868	4674
.3	2.0548	562$\overline{8}$	1.8192	6356	2.2473	6292	2.034$\overline{2}$	6951
.4	1.9811	0.741$\overline{6}$	1.7540	0.8376	2.1667	0.829$\overline{2}$	1.961$\overline{2}$	0.916$\overline{1}$
.5	1.887$\overline{6}$	0.913$\overline{0}$	1.671$\overline{2}$	1.031$\overline{2}$	2.0644	1.0208	1.8686	1.1278
.6	1.7752	1.075$\overline{3}$	1.5713	1.2145	1.9415	1.2023	1.757$\overline{4}$	1.328$\overline{3}$
.7	1.6451	1.228$\overline{8}$	1.4565	1.3856	1.7992	1.3717	1.628$\overline{6}$	1.515$\overline{5}$
.8	1.4985	1.3661	1.326$\overline{8}$	1.543$\overline{0}$	1.6389	1.527$\overline{5}$	1.483$\overline{5}$	1.6875
.9	1.3370	1.4917	1.183$\overline{8}$	1.6849	1.462$\overline{3}$	1.6679	1.323$\overline{6}$	1.842$\overline{7}$
1.0	1.162$\overline{2}$	1.6024	1.0289	1.8099	1.2710	1.7917	1.150$\overline{5}$	1.979$\overline{5}$
1.1	0.9756	1.6971	0.8638	1.9168	1.067$\overline{1}$	1.8976	0.965$\overline{9}$	2.096$\overline{5}$
1.2	.7794	1.774$\overline{9}$	.6900	2.0047	.8524	1.984$\overline{6}$	.771$\overline{6}$	2.1925
1.3	5754	1.8349	5094	2.0725	629$\overline{3}$	2.051$\overline{7}$	569$\overline{6}$	2.266$\overline{7}$
1.4	365$\overline{6}$	1.876$\overline{6}$	323$\overline{7}$	2.1196	3998	2.0983	3619	2.318$\overline{2}$
1.5	152$\overline{2}$	1.8996	1347	2.1455	1664	2.1239	1506	2.3465
$\tfrac{1}{2}\pi$	.0000	1.9043	0000	2.150$\overline{9}$	.0000	2.129$\overline{3}$	.0000	2.3524

Table III.

u	$\operatorname{gd} u$	$\theta°$	u	$\operatorname{gd} u$	$\theta°$	u	$\operatorname{gd} u$	$\theta°$
00	.0000	0.000	.60	.5669	32.483	1.50	1.1317	64.843
.02	020$\bar{0}$	1.146	.62	583$\bar{7}$	33.444	1.55	1.152$\bar{5}$	66.034
.04	040$\bar{0}$	2.291	.64	600$\bar{3}$	34.395	1.60	1.172$\bar{4}$	67.171
.06	060$\bar{0}$	3.436	.66	6167	35.336	1.65	1.1913	68.257
.08	0799	4.579	.68	6329	36.265	1.70	1.2094	69.294
.10	.0998	5.720	.70	.6489	37.183	1.75	1.226$\bar{7}$	70.284
.12	1197	6.859	.72	6648	38.091	1.80	1.243$\bar{2}$	71.228
.14	1395	7.995	.74	6804	38.987	1.85	1.258$\bar{9}$	72.128
.16	1593	9.128	.76	6958	39.872	1.90	1.273$\bar{9}$	72.987
.18	1790	10.258	.78	7111	40.746	1.95	1.2881	73.805
.20	.198$\bar{7}$	11.384	.80	.7261	41.608	2.00	1.3017	74.584
.22	218$\bar{3}$	12.505	.82	7410	42.460	2.10	1.3271	76.037
.24	2377	13.621	.84	755$\bar{7}$	43.299	2.20	1.350$\bar{1}$	77.354
.26	2571	14.732	.86	770$\bar{2}$	44.128	2.30	1.371$\bar{0}$	78.549
.28	2764	15.837	.88	7844	44.944	2.40	1.389$\bar{9}$	79.633
.30	.2956	16.937	.90	.798$\bar{5}$	45.750	2.50	1.407$\bar{0}$	80.615
.32	314$\bar{7}$	18.030	.92	8123	46.544	2.60	1.422$\bar{7}$	81.513
.34	3336	19.116	.94	826$\bar{0}$	47.326	2.70	1.436$\bar{6}$	82.310
.36	352$\bar{5}$	20.195	.96	8394	48.097	2.80	1.4493	83.040
.38	371$\bar{2}$	21.267	.98	8528	48.857	2.90	1.460$\bar{9}$	83.707
.40	.3897	22.331	1.00	.865$\bar{8}$	49.605	3.00	1.4713	84.301
.42	408$\bar{2}$	23.386	1.05	897$\bar{6}$	51.428	3.10	1.4808	84.841
.44	4264	24.434	1.10	9281	53.178	3.20	1.4894	85.336
.46	444$\bar{6}$	25.473	1.15	957$\bar{5}$	54.860	3.30	1.497$\bar{1}$	80.715
.48	462$\bar{6}$	26.503	1.20	985$\bar{7}$	56.476	3.40	1.504$\bar{1}$	86.177
.50	.4804	27.524	1.25	1.0127	58.026	3.50	1.5104	86.541
.52	4980	28.535	1.30	1.038$\bar{7}$	59.511	3.60	1.516$\bar{2}$	86.870
.54	5155	29.537	1.35	1.063$\bar{5}$	60.933	3.70	1.5214	87.168
.56	5328	30.529	1.40	1.087$\bar{3}$	62.295	3.80	1.526$\bar{1}$	87.437
.58	550$\bar{0}$	31.511	1.45	1.110$\bar{0}$	63.598	3.90	1.5303	87.681

Table IV.

u	$\operatorname{gd} u$	$\log \sinh u$	$\log \cosh u$	u	$\operatorname{gd} u$	$\log \sinh u$	$\log \cosh u$
4.0	1.534$\bar{2}$	1.4360	1.4363	5.5	1.5626	2.08758	2.0876$\bar{0}$
4.1	1.537$\bar{7}$	1.4795	1.4797	5.6	1.5634	2.13101	2.13103
4.2	1.5408	1.5229	1.5231	5.7	1.5641	2.17444	2.17445
4.3	1.543$\bar{7}$	1.5664	1.5665	5.8	1.5648	2.21787	2.21788
4.4	1.5462	1.6098	1.6099	5.9	1.5653	2.36130	2.26131
4.5	1.548$\bar{6}$	1.6532	1.6533	6.0	1.5658	2.30473	2.3047$\bar{4}$
4.6	1.550$\bar{7}$	1.6967	1.6968	6.2	1.5667	2.39159	2.3916$\bar{0}$
4.7	1.5526	1.7401	1.7402	6.4	1.567$\bar{5}$	2.47845	2.47846
4.8	1.5543	1.7836	1.7836	6.6	1.568$\bar{1}$	2.56531	2.56531
4.9	1.5559	1.8270	1.8270	6.8	1.568$\bar{6}$	2.65217	2.65217
5.0	1.5573	1.8704	1.870$\bar{5}$	7.0	1.569$\bar{0}$	2.73903	2.73903
5.1	1.5586	1.913$\bar{9}$	1.913$\bar{9}$	7.5	1.569$\bar{7}$	2.95618	3.95618
5.2	1.559$\bar{8}$	1.957$\bar{3}$	1.9573	8.0	1.570$\bar{1}$	3.1733$\bar{3}$	3.1733$\bar{3}$
5.3	1.5608	2.0007	2.0007	8.5	1.570$\bar{4}$	3.39047	3.39047
5.4	1.561$\bar{8}$	2.044$\bar{2}$	2.044$\bar{2}$	9.0	1.570$\bar{5}$	3.60762	3.60762
				∞	1.570$\bar{8}$	∞	∞

Article 40

Appendix.

40.1 Historical and Bibliographical.

What is probably the earliest suggestion of the analogy between the sector of the circle and that of the hyperbola is found in Newton's Principia (Bk. 2, prop. 8 et seq.) in connection with the solution of a dynamical problem. On the analytical side, the first hint of the modified sine and cosine is seen in Roger Cotes' Harmonica Mensurarum (1722), where he suggests the possibility of modifying the expression for the area of the prolate spheroid so as to give that of the oblate one, by a certain use of the operator $\sqrt{-1}$. The actual inventor of the hyperbolic trigonometry was Vincenzo Riccati, S.J. (Opuscula ad res Phys. et Math. pertinens, Bononiæ, 1757). He adopted the notation Sh.ϕ, Ch.ϕ for the hyperbolic functions, and Sc.ϕ, Cc.ϕ for the circular ones. He proved the addition theorem geometrically and derived a construction for the solution of a cubic equation. Soon after, Daviet de Foncenex showed how to interchange circular and hyperbolic functions by the use of $\sqrt{-1}$, and gave the analogue of De Moivre's theorem, the work resting more on analogy, however, than on clear definition (Reflex. sur les quant. imag., Miscel. Turin Soc., Tom. 1). Johann Heinrich Lambert systematized the subject, and gave the serial developments and the exponential expressions. He adopted the notation $\sinh u$, etc., and introduced the transcendent angle, now called the gudermanian, using it in computation and in the construction of tables (l. c. page 30). The important place occupied by Gudermann in the history of the subject is indicated on page 33.

The analogy of the circular and hyperbolic trigonometry naturally played a considerable part in the controversy regarding the doctrine of imaginaries, which occupied so much attention in the eighteenth century, and which gave birth to the modern theory of functions of the complex variable. In the growth of the general complex theory, the importance of the "singly periodic functions" became still clearer, and was gradually developed by such writers as Ferroni (Magnit. expon. log. et trig., Florence, 1782); Dirksen (Organon der tran. Anal.,

Berlin, 1845); Schellbach (Die einfach. period. funkt., Crelle, 1854); Ohm (Versuch eines volk. conseq. Syst. der Math., Nürnberg, 1855); Hoüel (Theor. des quant. complex, Paris, 1870). Many other writers have helped in systematizing and tabulating these functions, and in adapting them to a variety of applications. The following works may be especially mentioned: Gronau (Tafeln, 1862, Theor. und Anwend., 1865); Forti (Tavoli e teoria, 1870); Laisant (Essai, 1874); Gunther (Die Lehre ..., 1881). The last-named work contains a very full history and bibliography with numerous applications. Professor A. G. Greenhill, in various places in his writings, has shown the importance of both the direct and inverse hyperbolic functions, and has done much to popularize their use (see Diff. and Int. Calc., 1891). The following articles on fundamental conceptions should be noticed: Macfarlane, On the definitions of the trigonometric functions (Papers on Space Analysis, N. Y., 1894); Haskell, On the introduction of the notion of hyperbolic functions (Bull. N. Y. M. Soc., 1895). Attention has been called in Arts. 30 and 37 to the work of Arthur E. Kennelly in applying the hyperbolic complex theory to the plane vectors which present themselves in the theory of alternating currents; and his chart has been described on page 56 as a useful substitute for a numerical complex table (Proc. A. I. E. E., 1895). It may be worth mentioning in this connection that the present writer's complex table in Art. 39 is believed to be the only one of its kind for any function of the general argument $x + iy$.

40.2 Exponential Expressions as Definitions.

For those who wish to start with the exponential expressions as the definitions of $\sinh u$ and $\cosh u$, as indicated on page 27, it is here proposed to show how these definitions can be easily brought into direct geometrical relation with the hyperbolic sector in the form $\frac{x}{a} = \cosh \frac{S}{K}$, $\frac{y}{b} = \sinh \frac{S}{K}$, by making use of the identity $\cosh^2 u - \sinh^2 u = 1$, and the differential relations $d \cosh u = \sinh u\, du$, $d \sinh u = \cosh u\, du$, which are themselves immediate consequences of those exponential definitions. Let OA, the initial radius of the hyperbolic sector, be taken as axis of x, and its conjugate radius OB as axis of y; let $OA = a$, $OB = b$, angle $AOB = \omega$, and area of triangle $AOB = K$, then $K = \frac{1}{2}ab\sin\omega$. Let the coordinates of a point P on the hyperbola be x and y, then $\frac{x^2}{a^2} - \frac{y^2}{b^2} = 1$. Comparison of this equation with the identity $\cosh^2 u - \sinh^2 u = 1$ permits the two assumptions $\frac{x}{a} = \cosh u$ and $\frac{y}{b} = \sinh u$, wherein u is a single auxiliary variable; and it now remains to give a geometrical interpretation to u, and to prove that $u = \frac{S}{K}$, wherein S is the area of the sector OAP. Let the coordinates of a second point Q be $x + \Delta x$ and $y + \Delta y$, then the area of the triangle POQ is, by analytic geometry, $\frac{1}{2}(x\Delta y - y\Delta x)\sin\omega$. Now the sector POQ bears to the triangle POQ a ratio whose limit is unity, hence the differential of the sector S may be written $dS = \frac{1}{2}(xdy - ydx)\sin\omega = \frac{1}{2}ab\sin\omega(\cosh^2 u - \sinh^2 u)du = K du$. By integration $S = Ku$, hence $u = \frac{S}{K}$, the sectorial measure (p. 5); this establishes the fundamental geometrical relations $\frac{x}{a} = \cosh \frac{S}{K}$, $\frac{y}{b} = \sinh \frac{S}{K}$.

Printed in Poland
by Amazon Fulfillment
Poland Sp. z o.o., Wrocław